Lecture Notes in Mathematics　　1576

Editors:
A. Dold, Heidelberg
B. Eckmann, Zürich
F. Takens, Groningen

Kazuaki Kitahara

Spaces of Approximating Functions with Haar-like Conditions

Springer-Verlag

Berlin Heidelberg New York
London Paris Tokyo
Hong Kong Barcelona
Budapest

Author

Kazuaki Kitahara
Department of Mathematics
Faculty of Education
Saga University
Saga 840, Japan

Mathematics Subject Classification (1991): 41A50, 41A30

ISBN 3-540-57974-5 Springer-Verlag Berlin Heidelberg New York
ISBN 0-387-57974-5 Springer-Verlag New York Berlin Heidelberg

CIP-Data applied for

© Springer-Verlag Berlin Heidelberg 1994
Printed in Germany

Typesetting: Camera-ready by author/editor
SPIN: 10130001   46/3140-543210 - Printed on acid-free paper

To Junko and Toshiya

Preface

Let E be a function space with a norm $\|\cdot\|$ and let G be a finite dimensional subspace of E. Then it is one of the principal themes in approximation theory to study the following problems: For each $f \in E$,

find $\tilde{f} \in G$ such that $E_G(f) = \|f - \tilde{f}\| = \inf_{g \in G} \|f - g\|$

and

estimate $E_G(f)$.

G is called an approximating space and $\tilde{f}$ is said to be a best approximation to f from G. If G is chosen in the manner so that $E_G(f)$ is as small as possible and so that functions in G are easy to handle, then G is a good approximating space.

For example, in $C[a, b]$ (=the space of all real-valued continuous functions on $[a, b]$) with the supremum norm, spaces of polynomials with degree at most n and spaces of continuous and piecewise linear functions with fixed knots are suitable for good approximating spaces. Čebyšev (or Haar) spaces and weak Čebyšev spaces are generalizations of these two spaces and play a central part when considering the above problems. In fact, properties, characterizations and generalizations of Čebyšev and weak Čebyšev spaces have been deeply studied during this century. Now, the theory of these spaces has matured.

In this book, as approximating spaces, we shall introduce Haar-like spaces, which are Haar and weak Čebyšev spaces under special conditions. And we shall study topics of subclasses of Haar-like spaces rather than general properties of Haar-like spaces, that is, classes of Čebyšev or weak Čebyšev spaces, spaces of vector-valued monotone increasing or convex functions and spaces of step functions.

Contents are mostly new results and rewritings of the following papers, 13, 14, 15, 16, 17 (Chapter 2), 7, 8, 9(Chapter 3), 17, 18 (Chapter 4), 4, 5 (Chapter 5), 2 (Appendix 1), where each number is its reference number. In Chapter 1, Haar-like spaces are defined and several examples of Haar-like spaces are given. In Chapter 2 and 3, for Čebyšev and Čebyšev-like spaces, we are concerned with characterizations, derivative spaces, separated representations, adjoined functions and best L^1-approximations. In Chapter 4, in a space of vector-valued functions of bounded variation, we consider best approximations by monotone increasing or convex functions. In Chapter 5, approximation by step functions is studied. In connection with Chapter 5, Dirichlet tilings and a certain property of the finite decomposition of a set are stated in Appendix 1 and 2, respectively. The readers can see further summary in the Introduction to each chapter.

I would like to express my heartfelt gratitude to emeritus Prof. Kiyoshi Iseki at Naruto Education University and Prof. Shirô Ogawa at Kwansei Gakuin University who taught me topological vector spaces and approximation theory and who have given me constant encouragement with high degree. I am indebted to Katsumi Tanaka, Takakazu Yamamoto and Hiroaki Katsutani for their help in preparing this manuscript, to Donna L. DeWick for checking the style of this manuscript and to the editors and staff of Springer-Verlag for their able cooperation.

Finally I am also grateful for the constant heartwarming encouragement from my parents, Torao Kitahara and Yasue Kitahara.

Kazuaki Kitahara

Saga, Japan

January, 1994

Contents

Chapter 1 Preliminaries

Chapter 2 Characterizations of Approximating Spaces of $C[a,b]$ or $C_0(Q)$

Chapter 3 Some Topics of Haar-like Spaces of $F[a,b]$

Chapter 4 Approximation by Vector-Valued Monotone Incresing or Convex Functions

Chapter 5 Approximation by Step Functions

Appendix 1

Appendix 2

Chapter 1 Preliminaries

1.1 Introduction

Before stating the purpose of this chapter, we shall introduce a concrete problem in approximation theory. Let $C[a, b]$ be the space of all real-valued functions on a compact real interval $[a, b]$. $C[a, b]$ is endowed with the supremum norm $\| \cdot \|$,i.e., $\|f\| = \sup_{x \in [a,b]} |f(x)|$ for all $f \in C[a, b]$. For a finite dimensional subspace G of $C[a, b]$, we consider the following problem: For a given $f \in C[a, b]$,

$$\text{find } \tilde{f} \in G \text{ such that } \|f - \tilde{f}\| = \inf_{g \in G} \|f - g\|,$$

in other words,

$$\text{find the best approximations } \tilde{f} \text{ to } f \text{ from } G.$$

G is a space of approximating functions. It is well known that spaces spanned by the following systems $\{u_i\}_{i=1}^n$ in $C[a, b]$ are of good use to treat this problem. One is a generalization of systems of spline functions and the other is a generalization of systems of polynomials.

(1) $\{u_i\}_{i=1}^n$ is a system such that, for any n distinct points $(a \leq)x_1 < \ldots < x_n(\leq b)$,

$$\det \begin{pmatrix} \sigma u_1(x_1) & \cdots & u_n(x_1) \\ \vdots & \cdots & \vdots \\ \sigma u_1(x_n) & \cdots & u_n(x_n) \end{pmatrix} \geq 0,$$

where $\sigma = 1$ or -1.

(2) $\{u_i\}_{i=1}^n$ is a system such that, for any n distinct points $(a \leq)x_1 < \ldots < x_n(\leq b)$,

$$\det \begin{pmatrix} u_1(x_1) & \cdots & u_n(x_1) \\ \vdots & \cdots & \vdots \\ u_1(x_n) & \cdots & u_n(x_n) \end{pmatrix} \neq 0.$$

Systems of (1) are called weak Tchebycheff, weak Čebyšev or WT-systems and those of (2) are called Haar, Tchebycheff, Čebyšev or T-systems.(Unified terms are not used for these systems.) One useful property of WT and T-systems is its marked characterization of best approximations from these spaces. When spaces spanned by WT and T-systems are approximating spaces, best approximations in the above problem are chracterized as follows.

Theorem 1.1.1. *Let $\{u_i\}_{i=1}^n$ be a WT-system in $C[a, b]$ and let G be the space spanned by $\{u_i\}_{i=1}^n$. For a given $f \in C[a, b]$, if there exist an $\tilde{f} \in G$ and $n + 1$ points*

1

$(a \leq) \ x_1 < \cdots < x_{n+1} (\leq b)$ such that $\|f - \tilde{f}\| = \sigma(-1)^i(f(x_i) - \tilde{f}(x_i))$ $(\sigma = 1$ or $-1)$, $i = 1, \ldots, n+1$, then $\tilde{f}$ is a best approximation to f from G. Furthermore, each $f \in C[a, b]$ has a best approximation from G satisfying this condition.

Theorem 1.1.2. Let $\{u_i\}_{i=1}^n$ be a T-system in $C[a, b]$ and let G be the space spanned by $\{u_i\}_{i=1}^n$. Each $f \in C[a, b]$ has a unique best approximation from G satisfying the condition stated in Theorem 1.1.1.

These systems play an important role not only in characterizations of best approximations, but also in interpolation methods, moment spaces, totally positive kernels and so on.

Some generalizations of WT and T-systems have already been defined and investigated. (e.g. Ault, Deutsch, Morris and Olson[1], Garkavi[4] e.t.c.) In this chapter, we will define other types of generalized WT and T-systems. Furthermore, we mainly study these systems in the following chapters.

1.2 Definitions

We begin by giving definitions of Haar-like conditions.

Definition 1. Let E be a real vector space and let E^* be the algebraic dual space of E, i.e., the space of all real-valued linear functionals on E. For a given positive integer n, let $\{a_i\}_{i=1}^n$ be n linearly independent elements in E and let $\mathcal{F}$ be a subset of $(E^*)^n \ (= \overbrace{E^* \times \cdots \times E^*}^{n})$.

(1) If, for any element $(x_1, \ldots, x_n) \in \mathcal{F}$, the n-th order determinant

$$D \begin{pmatrix} a_1 & \cdots & a_n \\ x_1 & \cdots & x_n \end{pmatrix} := \det \begin{pmatrix} x_1(a_1) & \cdots & x_1(a_n) \\ \vdots & \cdots & \vdots \\ x_n(a_1) & \cdots & x_n(a_n) \end{pmatrix} \neq 0,$$

then it is said that $\{a_i\}_{i=1}^n$ satisfies H condition with $\mathcal{F}$ or is an H-system with $\mathcal{F}$(abbreviated $H_{\mathcal{F}}$-system). The space $[a_1, \ldots, a_n]$ spanned by elements of an $H_{\mathcal{F}}$-system $\{a_i\}_{i=1}^n$ is called an $H_{\mathcal{F}}$-space.

(2) If, for any element $(x_1, \ldots, x_n) \in \mathcal{F}$,

$$D \begin{pmatrix} \sigma a_1 & \cdots & a_n \\ x_1 & \cdots & x_n \end{pmatrix} > 0, \quad (\text{resp.} \geq 0)$$

2

where $\sigma = 1$ or -1, then it is said that $\{a_i\}_{i=1}^n$ satisfies T *condition* (resp. WT *condition*) *with* $\mathcal{F}$ or is a $T_{\mathcal{F}}$-*system* (resp. $WT_{\mathcal{F}}$-*system*). And, we call the space spanned by a $T_{\mathcal{F}}$-system (resp. $WT_{\mathcal{F}}$-system) a $T_{\mathcal{F}}$-*space* (resp. $WT_{\mathcal{F}}$-*space*).

(3) Let $\mathcal{F}_k = \{(x_1, \ldots, x_k) \mid (x_1, \ldots, x_n) \in \mathcal{F}\}$ for each k, $1 \leq k \leq n$. If $\{a_i\}_{i=1}^n$ is a system such that, for each k, $1 \leq k \leq n$, $\{a_i\}_{i=1}^k$ is an $H_{\mathcal{F}_k}$-system (resp. $T_{\mathcal{F}_k}$, $WT_{\mathcal{F}_k}$-system), then it is called a *complete* $H_{\mathcal{F}}$-*system* (resp. *complete* $T_{\mathcal{F}}$, *complete* $WT_{\mathcal{F}}$-*system*).

(4) For our convenience, when we take these three conditions in (1), (2) and (3) together, we call them *Haar-like conditions*. Analogously, we use the terms *Haar-like systems* and *Haar-like spaces*.

Now we add two approximating spaces in a normed space.

Definition 2. Let E be a normed space with a norm $\| \cdot \|$ and let M be a subset of E.

(1) For a given $x \in E$, if there exists an $x_0 \in M$ such that $\|x - x_0\| = \inf_{y \in M} \|x - y\|$, then x_0 is called a *best approximation to* x *from* M or simply a *best approximation to* x. The set of all best approximations to x from M is denoted by $P_M(x)$.

(2) For a subspace G of E, let $U_G = \{x \mid x \in E, P_G(x) \text{ is singleton }\}$. If $U_G = E$, G is called a *C-space*, and if $E - U_G$ is a set of the first category in E, then G is said to be an *AC-space*.

Remark 1. (1) $\{a_i\}_{i=1}^n$ is a $T_{\mathcal{F}}$-system if and only if it is a $WT_{\mathcal{F}}$ and $H_{\mathcal{F}}$-system. But every $WT_{\mathcal{F}}$ or $H_{\mathcal{F}}$-system is not always a $T_{\mathcal{F}}$-system.

(2) As for completeness of Haar-like systems, we can define a more specialized system such as order complete or Descartes systems (see Nürnberger[8; p.15]). But these systems are not addressed in this book.

(3) Let $\{a_i\}_{i=1}^n$ be a system of E and let $\mathcal{F}$ be a subset of $(E^*)^n$. For any $(x_1, \ldots, x_n)$ $\in \mathcal{F}$ and each k, $1 \leq k \leq n$, $x_i^{(k)}$ is a linear functional on E such that $x_i^{(k)}(a_j) = x_i(a_j)$ if $1 \leq i, j \leq k$, $x_i^{(k)}(a_j) = 1$ if $i = j > k$, $x_i^{(k)}(a_j) = 0$ otherwise. If, for each $\{a_i\}_{i=1}^n$, we put $\mathcal{F}' = \{(x_1^{(k)}, \ldots, x_n^{(k)}) \mid (x_1, \ldots, x_n) \in \mathcal{F}, 1 \leq k \leq n\}$, then $\{a_i\}_{i=1}^n$ is a complete $H_{\mathcal{F}}$-system if and only if it is an $H_{\mathcal{F}'}$-system.

(4) The definition of AC-spaces is introduced by Stečkin[12].

(5) Best approximations in general normed spaces are studied in detail in Singer[11].

From Definition 1, the following statement immediately follows.

3

Proposition 1.2.1. *Let E be a real normed space and let E' be the topological dual space of E, i.e., the space of all real-valued continuous linear functionals on E. Let $\mathcal{F}$ be a connected subset of $(E')^n$, where E' is endowed with the weak topology $\sigma(E', E)$. Then if $\{a_i\}_{i=1}^n$ is an $H_{\mathcal{F}}$-system in E, it is a $T_{\mathcal{F}}$-system.*

1.3 Examples of Haar-like Spaces

We give some function spaces and examples of Haar-like subspaces, which are studied in the following chapters.

1. Let E be a real normed space and let E' be the topological dual space of E. $S_{E'}$ denotes the closed unit ball in E' and the set of extreme points of $S_{E'}$ is denoted by $extS_{E'}$. An n-dimensional subspace M of E is called an *interpolating subspace* if, for any n linearly independent functionals $x_1, \ldots, x_n$ in $extS_{E'}$ and any n real scalars $c_1, \ldots, c_n$, there is a unique element $a \in M$ such that $x_i(a) = c_i$ for $i = 1, \ldots, n$. Ault, Deutsch, Morris and Olson[1] gave the definition of interpolating subspaces and studied best approximations from interpolating subspaces in detail.

As a subset $\mathcal{F}$ of $(E')^n$, setting $\mathcal{F} = \{(x_1, \ldots, x_n) | \ x_1, \ldots, x_n$ are linearly independent functionals in $extS_{E'}\}$, we can consider that every interpolating subspace is an $H_{\mathcal{F}}$-space.

Interpolating spaces are closely related with C-spaces.

Proposition 1.3.1. (Ault, Deutsch, Morris and Olson[1; Theorem 2.2]) *Let M be a finite dimensional subspace of a real normed space. If M is an interpolating space, then it is a C-space.*

2. For a set X, $F(X)$ denotes the space of all real-valued functions on X. We easily see that each point x in X is a linear functional on $F(X)$ such that $x(f) = f(x)$ for all $f \in F(X)$. Let $\{A_i\}_{i=1}^n$ be an n-decomposition of X, i.e., $A_i \cap A_j = \emptyset$ for $i \neq j$ and $\cup_{i=1}^n A_i = X$, and let $\chi_{A_i}(x)$, $i = 1, \ldots n$, be the characteristic function of A_i. Then $[\chi_{A_1}, \ldots, \chi_{A_n}]$ is an n-dimensional subspace of $F(X)$ which consists of step functions on X.

If we consider a subset of $(E^*)^n$, $\mathcal{F} = \{(x_1, \ldots, x_n) | x_i \in A_i, i = 1, \ldots, n\}$, $\{\chi_{A_i}\}_{i=1}^n$ can be regarded as a $T_{\mathcal{F}}$-system.

4

Problems of approximation by step functions will be studied in Chapter 5.

3. Let T be a partially ordered set with an order $\leq$. As a subset of $(F(T)^*)^n$, setting $\mathcal{P} = \{(x_1, \ldots, x_n) \mid x_1 < \cdots < x_n, x_i \in T, i = 1, \ldots, n\}$ ($x < y$ means that $x \leq y$ and $x \neq y$.), we can consider $H_{\mathcal{P}}$, $T_{\mathcal{P}}$ and $WT_{\mathcal{P}}$-systems. When T is a linearly ordered set, Zielke[14] studied many properties of $H_{\mathcal{P}}$, $T_{\mathcal{P}}$ and $WT_{\mathcal{P}}$-systems in $F(T)$ (Zielke[14] uses other terminologies for these systems) which consist of discontinuous functions.

From Lemma 1.2 and Lemma 4.1 in Zielke[14], we have

Proposition 1.3.2. *Let T be a linearly ordered set containing at least $n+1$ points and $\{u_i\}_{i=1}^n$ be a system in $F(T)$.*

(1) *$\{u_i\}_{i=1}^n$ is an $H_{\mathcal{P}}$-system if and only if any $u \in [u_1, \ldots, u_n] - \{0\}$ has at most $n-1$ zeros.*

(2) *$\{u_i\}_{i=1}^n$ is a $WT_{\mathcal{P}}$-system if and only if no $u \in [u_1, \ldots, u_n]$ has an alternation of length $n+1$, i.e., there do not exist $n+1$ points $x_1 < \cdots < x_{n+1}$ in T such that $(-1)^i u(x_i)$ is positive for $i = 1, \ldots, n+1$ or negative for $i = 1, \ldots, n+1$.*

A necessary and sufficient condition of $T_{\mathcal{P}}$-systems immediately follows from the conditions in (1) and (2) in Proposition 1.3.2. Clearly, Tchebycheff systems (resp. weak Tchebycheff systems) in 1.1 are identical with $T_{\mathcal{P}}$-systems (resp. $WT_{\mathcal{P}}$-systems).

4. For a Hausdorff topological space X, $C(X)$ denotes the space of all real-valued continuous functions on X and $C_0(X)$ denotes the subspace of $C(X)$ which consists of functions f such that $\{x \in X \mid |f(x)| \geq \epsilon\}$ is compact for each $\epsilon > 0$.

In this book, we mainly treat X as a subset of the real line R. In particular, when X is a nondegenerate compact interval $[a, b]$, we use the notation $C[a, b]$ instead of $C([a, b])$. And as another function space on $[a, b]$, we denote the space of all real-valued Lebesgue integrable functions on $[a, b]$ by $L^1[a, b]$.

As is stated in 1.1, properties of $H_{\mathcal{P}}$, $T_{\mathcal{P}}$ and $WT_{\mathcal{P}}$-systems in $C[a, b]$ (Other terminologies are used for these systems) have been profoundly studied. We can observe many good properties of these systems in books and journals related to approximation theory.(e.g. Cheney[2], Davis[3], Karlin and Studden[5], Lorentz[7], Watson[13] etc.)

Now we introduce other types of systems in $C[a, b]$ or $L^1[a, b]$. Let $\mathcal{S}$ be the set of all nondegenerate subintervals of $[a, b]$. For $I, J \in \mathcal{S}$, if the interior points of I equal those of J, we write $I = J$, and if $I \cap J$ has no interior points and $x \leq y$ for all $x \in I$

5

and $y \in J$, then it is denoted by $I < J$ for this relation. By this, $(\mathcal{S}, \leq)$ is a partially ordered set, and $\leq$ means $=$ or $<$. For each $I \in \mathcal{S}$, we define a linear functional u_I on $C[a, b]$ or $L^1[a, b]$ such that $u_I(f) = \int_I f(x)dx$ for all $f \in C[a, b]$ or $L^1[a, b]$ and let $\mathcal{I} = \{ (u_{I_1}, \ldots, u_{I_n}) \mid I_i \in \mathcal{S}, \ i = 1, \ldots, n, \ I_1 < \cdots < I_n \}$. Then we can consider $H_{\mathcal{I}}$, $T_{\mathcal{I}}$ and $WT_{\mathcal{I}}$-systems in $C[a, b]$ or $L^1[a, b]$.

The readers can easily obtain the following proposition from the above definition.

Proposition 1.3.3. *Let $\{u_i\}_{i=1}^n$ be a system in $C[a, b]$ or $L^1[a, b]$. $\{u_i\}_{i=1}^n$ is an $H_{\mathcal{I}}$-system if and only if it is a $T_{\mathcal{I}}$-system.*

Shi[10] proposes a variation of L approximation which maintains almost all of the Chebyshev theory and considers best approximations by $H_{\mathcal{I}}$-systems ($= QT$-systems in Shi[10]). Basic properties of $H_{\mathcal{I}}$-systems in $C[a, b]$ are studied in Kitahara[6]. Further properties of $H_{\mathcal{I}}$ and $WT_{\mathcal{I}}$-systems will be investigated in Chapter 2 and 3.

5. Let $C^{n-1}[a, b]$ $(n \geq 1)$ be the subspace of $C[a, b]$ which consists of $n - 1$ times continuously differentiable functions and let $\{u_i\}_{i=1}^n$ be a system in $C^{n-1}[a, b]$. For any $\mathbf{x} = (x_1, \ldots, x_n)$ with $a \leq x_1 \leq \ldots \leq x_n \leq b$, we define the following linear functionals $z_i^{\mathbf{x}}$ on $C^{n-1}[a, b]$:

$$z_i^{\mathbf{x}}(f) = f^{(k_i)}(x_i) \quad \text{for } f \in C^{n-1}[a, b], \ i = 1, \ldots, n,$$

where $k_i = \max \{j \mid j \text{ is a nonnegative integer and } x_i = \cdots = x_{i-j}\}, \ i = 1, \ldots, n$. If $\{u_i\}_{i=1}^n$ satisfies the condition that

$$D \begin{pmatrix} \sigma u_1 & \cdots & u_n \\ z_1^{\mathbf{x}} & \cdots & z_n^{\mathbf{x}} \end{pmatrix} > 0$$

for all $\mathbf{x} = (x_1, \ldots, x_n)$ with $a \leq x_1 \leq \ldots \leq x_n \leq b$, then $\{u_i\}_{i=1}^n$ is called an extended Tchebycheff system of order n or simply an extended Tchebycheff system (see p.6 in Karlin and Studden[5] and p.4 in Nürnberger[8]). Setting $\tilde{\mathcal{P}} = \{(z_1^{\mathbf{x}}, \ldots, z_n^{\mathbf{x}}) \mid a \leq x_1 \leq \ldots \leq x_n \leq b\}$, every extended Tchebycheff system is identical with a $T_{\tilde{\mathcal{P}}}$-system. In particular, if $\mathbf{x} = (t, \ldots, t)$, $t \in [a, b]$, $D \begin{pmatrix} \sigma u_1 & \cdots & u_n \\ z_1^{\mathbf{x}} & \cdots & z_n^{\mathbf{x}} \end{pmatrix}$ denotes Wronskian determinants $W(u_1, \ldots, u_n)(t)$ of $\{u_i\}_{i=1}^n$ at $t \in [a, b]$.

6. We give an example of generalized convex functions in $F[a, b]$ by using $WT_{\mathcal{P}}$-condition. Let $\mathcal{U} = \{u_i\}_{i=1}^n$ be a system in $F[a, b]$ consisting of linearly independent functions. If f is a function in $F[a, b]$ such that $\{u_i\}_{i=1}^n \cup \{f\}$ is a $WT_{\mathcal{P}}$-system, then

6

f is called a $\mathcal{U}$-convex function. When $\mathcal{U} = \{1, x, x^2, \ldots, x^{n-1}\}$, $\mathcal{U}$-convex functions are called n-convex functions.(see Roberts and Varberg[9], Zwick[15])

Analogously, we can consider vector-valued $\mathcal{U}$-convex functions. Let $\mathcal{U} = \{u_i\}_{i=1}^n$ be a system in $F[a, b]$ consisting of linearly independent real-valued functions, and let E be an ordered real vector space. If f is an E-valued function on $[a, b]$ such that, for any $(x_1, \ldots, x_{n+1}) \in \mathcal{P}$, the $n + 1$-th order determinant

$$
\det \begin{pmatrix} u_1(x_1) & \cdots & u_1(x_{n+1}) \\ \vdots & \cdots & \vdots \\ u_n(x_1) & \cdots & u_n(x_{n+1}) \\ f(x_1) & \cdots & f(x_{n+1}) \end{pmatrix} \geq 0,
$$

then we call f an E-valued $\mathcal{U}$-convex function. In this book, we do not study properties of vector-valued $\mathcal{U}$-convex functions, but we will treat approximation by vector-valued 1-convex or 2-convex functions in Chapter 4.

Remark 2. (1) Let $\mathcal{U}$ be an extended complete Tchebycheff system. Karlin and Studden studied properties of the set (= cone) of all real-valued $\mathcal{U}$-convex functions in depth(see Karlin and Studden[5; Chapter XI]).

(2) Any system in a real vector space can be an $H_{\mathcal{F}}$-system or $WT_{\mathcal{F}}$-system for some subset $\mathcal{F} \subset (E^*)^n$. Hence, it is important to consider Haar-like systems under ideal subsets of $(E^*)^n$.

(3) We shall use the terminologies introduced here throughout this book.

1.4 Problem

1. Let M be a finite dimensional subspace of a real normed space $(E, \|\cdot\|)$. If M is an $H_{\mathcal{F}}$-space, where $\mathcal{F}$ is a subset of $(E')^n$ in example 1 in 1.3, then M is a C-space. Similarly is there any subset $\mathcal{G}$ of $(E')^n$ such that an $H_{\mathcal{G}}$-space is an AC-space ?

Chapter 2 Characterizations of Approximating
Spaces of $C[a, b]$ or $C_0(Q)$

2.1 Introduction

Let Q be a locally compact subset of R and let $C_0(Q)$ and $C[a, b]$ be the function spaces defined in 1.3. $C_0(Q)$ and $C[a, b]$ are endowed with the supremum norm $\|\cdot\|$, i.e., $\|f\| = \sup_{x \in Q(x \in [a,b])} |f(x)|$ for each $f \in C_0(Q)(C[a, b])$. Let $\mathcal{P}$ and $\mathcal{I}$ be the subsets of $(C_0(Q)^*)^n$ or $(C[a, b]^*)^n$ defined in 1.3.

In this chapter, we introduce known characterizations of approximating spaces of $C_0(Q)$ or $C[a, b]$ and show other types of characterizations of these spaces.

In 2.2, basic properties of $H_\mathcal{P}$, $WT_\mathcal{P}$, C and AC-spaces in $C_0(Q)$ are observed. In 2.3, we review characterizations of $T_\mathcal{P}(= H_\mathcal{P})$, $WT_\mathcal{P}$ and $H_\mathcal{I}(= T_\mathcal{I})$-spaces of $C[a, b]$. These are stated in terms of sets defined by best approximations. In general, all results in this section can not hold in $C_0(Q)$. In 2.4, material similar to 2.3 is introduced and some of the results in 2.3 are extended to $C_0(Q)$.

In 2.5 and 2.6, we consider different types of characterizations from those in 2.3 and 2.4. In 2.5, we show a characterization of $H_\mathcal{P}(= T_\mathcal{P})$-spaces of $C(R)$ in terms of appropriate decompositions of R^2. In 2.6, using the nonexistence theorem of best approximations from a closed subspace, we give a chracterization of a space spanned by an infinite complete $T_\mathcal{P}(= H_\mathcal{P})$-system. Finally, some problems related to these topics are stated in 2.7.

2.2 Approximating Spaces of $C_0(Q)$

Let Q be a locally compact subset of R and let $C_0(Q)$ be the Banach space defined in 2.1.

Let us recall that, for an n-dimensional subspace G of $C_0(Q)$, if $U_G = C_0(Q)$, G is said to be a C-space and if $C_0(Q) - U_G$ is a set of the first category in $C_0(Q)$, then G is called an AC-space.

By Theorem 3.2 in Ault, Deutsch, Morris and Olson[1], we state

Proposition 2.2.1. *Let G be an n-dimensional subspace in $C_0(Q)$. The following statements are equivalent:*

8

(1) G is an H_P-space.

(2) For any n distinct points $x_1, \ldots, x_n$ in Q and any real numbers $c_1, \ldots, c_n$, there exists an $f \in G$ such that $f(x_i) = c_i$, $i = 1, \ldots, n$.

(3) G is a C-space.

(4) G is an interpolating subspace.

The equivalence of (1) and (2) follows from a straightfoward application of the definition of H_P-spaces. The equivalence of (1),(3) and (4) follows from Theorem 3.2 in Ault, Deutsch, Morris and Olson [1].

As a special case of Proposition 1.3.2, we state

Proposition 2.2.2. Let G be an n-dimensional subspace in $C_0(Q)$. The following statements are equivalent.

(1) G is a WT_P-space.

(2) No $f \in G$ has an alternation of length $n + 1$.

For every subset A of Q, we associate the number $N_n(A)$ equal to the number of points of A, if this number does not exceed n, and equal to n otherwise.

If G is a finite dimensional AC-space, then we have

Proposition 2.2.3. Let Q be a locally compact subset of R which contains at least n points and let G be an n-dimensional subspace of $C_0(Q)$. The following statements are equivalent:

(1) G is an AC-space.

(2) On each open subset $O \subset Q$, at most $n - N_n(O)$ linearly independent functions in G can vanish identically.

(3) There exists a dense subset $\mathcal{P}'$ of $\mathcal{P}$ such that G is an $H_{\mathcal{P}'}$-space.

Proof.

(1) $\leftrightarrow$ (2) Since $C_0(Q)$ is separable, the equivalence of (1) and (2) follows from the same proof of Theorem 1 in Garkavi[7].

(2) $\rightarrow$ (3) Suppose that $f_1, \cdots, f_n$ is a basis for G. It is sufficient to show a dense subset $\mathcal{P}'$ of $\mathcal{P}$ such that, for each $(x_1, \ldots, x_n) \in \mathcal{P}'$, $D \begin{pmatrix} f_1 & \cdots & f_n \\ x_1 & \cdots & x_n \end{pmatrix} \neq 0$.

Let $x_1, \ldots, x_n$ be any n distinct points in Q with $x_1 < \cdots < x_n$ and let O_i, $i = 1, \ldots, n$, be any open neighbourhood of x_i such that $O_i \cap O_j = \emptyset$ for $i \neq j$. Without loss of generality, we may assume that each O_i, $i = 1, \ldots, n$, is a one point set or an

9

infinite points set. Suppose that $O_1, \ldots, O_k$ are one point sets and $O_{k+1}, \ldots, O_n$ are infinite points sets. From (2), we have

(2.2.1) $\dim Span\{(f_1(x), \ldots, f_n(x)) | x \in O_1 \cup \cdots \cup O_k\} = k$

(2.2.2) $\dim Span\{(f_1(x), \ldots, f_n(x)) | x \in O_i\} = n, i = k+1, \ldots, n.$

By (2.2.1) and (2.2.2), there exist points $y_i \in O_i$, $i = k+1, \ldots, n$, such that

$$\dim Span\{(f_1(x), \ldots, f_n(x)) | x \in O_1 \cup \cdots \cup O_k \cup \{y_{k+1}, \ldots, y_n\}\} = n.$$

This implies that $D \begin{pmatrix} f_1 & \cdots & & \cdots & f_n \\ x_1 & \cdots & x_k & y_{k+1} & \cdots & y_n \end{pmatrix} \neq 0$. Hence, a dense subset $\mathcal{P}'$ of $\mathcal{P}$ exists.

(3) $\rightarrow$ (2). Let O be any open subset of Q. Suppose that O consists of $k (\leq n-1)$ points $x_1, \ldots, x_k$, i.e., O is a set of k isolated points. Then we take $n - k$ distinct points $y_1, \ldots, y_{n-k}$ in $Q - O$ and $n - k$ neighbourhoods U_i of y_i, $i = 1, \ldots, n - k$, such that $U_i \cap U_j = \emptyset$ for $i \neq j$. By an application of (3), for these n points $x_1, \ldots, x_k, y_1, \ldots, y_{n-k}$, there are $n - k$ points z_i in U_i, $i = 1, \ldots, n - k$, such that $D \begin{pmatrix} f_1 & \cdots & & \cdots & f_n \\ x_1 & \cdots & x_k & z_1 & \cdots & z_{n-k} \end{pmatrix} \neq 0$. This means that $\dim Span\{(f_1(x), \ldots, f_n(x)) | x \in \{x_1, \ldots, x_k\}\} = k$.

Suppose that O contains at least n points. By (3), we easily find n distinct points $x_1, \ldots, x_n$ in O satisfying $D \begin{pmatrix} f_1 & \cdots & f_n \\ x_1 & \cdots & x_n \end{pmatrix} \neq 0$. Hence, (2) follows from this.

2.3 Characterizations of Approximating Spaces of C[a,b]

In this section, we review characterizations of approximating spaces of $C[a, b]$. Before doing this, we state necessary and sufficient conditions for a system to be an H_I ($= T_I$)-system.

Proposition 2.3.1. *Let* $f_1, \ldots, f_n$ *be linearly independent functions in* $C[a, b]$. *The following statements are equivalent:*

(1) $\{f_i\}_{i=1}^n$ *is an* H_I-*system.*

(2) $\{f_i\}_{i=1}^n$ *is a* WT_P-*system and, for any* $f \in [f_1, \ldots, f_n] - \{0\}$, $Z(f) = \{x | x \in [a, b], f(x) = 0\}$ *is nowhere dense in* $[a, b]$.

(3) $D \begin{pmatrix} \sigma f_1 & \cdots & f_n \\ x_1 & \cdots & x_n \end{pmatrix} \geq 0$ *for all* $(x_1, \ldots, x_n) \in \mathcal{P}$, *where* σ *is a constant 1 or* -1, *and* $\mathcal{P}' = \{(x_1, \ldots, x_n) | (x_1, \ldots, x_n) \in \mathcal{P}, D \begin{pmatrix} \sigma f_1 & \cdots & f_n \\ x_1 & \cdots & x_n \end{pmatrix} > 0\}$ *is dense in* $\mathcal{P}$.

(4) $V = \{x \mid x \in [a, b], f_i(x) = 0, i = 1, \ldots, n\}$ *is nowhere dense in* $[a, b]$ *and* $[f_1, \ldots, f_n]|_{(a,b)-V}$ *is an* H_P-*space,i.e., on* $(a, b) - V$, *every* $f \in [f_1, \ldots, f_n] - \{0\}$ *has at most* $n - 1$ *zeros.*

(5) $[f_1, \ldots, f_n]$ *is a* WT_P *and AC-space.*

Proof. (1) $\rightarrow$ (2). Assume that $\{f_i\}_{i=1}^n$ is not a WT_P-system. There is a function $f \in [f_1, \ldots, f_n] - \{0\}$ which satisfies, for some $(a <)x_1 < \cdots < x_{n+1}(< b)$,

$$(2.3.1) \qquad f(x_i)f(x_{i+1}) < 0, \quad i = 1, \ldots, n.$$

By (2.3.1), there exist disjoint subintervals $J_i \subset [x_i, x_{i+1}]$, $i = 1, \ldots, n$, satisfying

$$(2.3.2) \qquad \int_{J_i} f dx = 0, \quad i = 1, \ldots, n.$$

But (2.3.2) contradicts the definition of H_I-systems. Next, suppose that for some $g \in [f_1, \ldots, f_n] - \{0\}$, the set of interior points of $Z(g)$ is not empty. By this assumption, we find disjoint subintervals I_i, $i = 1, \ldots, n$, such that g vanishes identically on each I_i. This also contradicts (1).

(2) $\rightarrow$ (1). Suppose that $\{f_i\}_{i=1}^n$ is not an H_I-system. Then, for some $f \in [f_1, \ldots, f_n] - \{0\}$ and some disjoint subintervals J_i, $i = 1, \ldots, n$, of $[a, b]$,

$$(2.3.3) \qquad \int_{J_i} f dx = 0, \quad i = 1, \ldots, n.$$

Since f does not vanish identically on each J_i, from the continuity of f and (2.3.3), we obtain

$$f(t_i)f(s_i) < 0 \quad \text{for some } t_i, s_i \in I_i, \quad i = 1, \ldots, n.$$

This contradicts the fact that $\{f_i\}_{i=1}^n$ is a WT_P-system.

(3) $\rightarrow$ (1). For $I_1 < \cdots < I_n$, applying problem 68, p.61 in Pólya and Szegö[19], we see that

$$
\begin{aligned}
\det(\int_{I_j} f_i(x)dx)_{i,j=1}^n &= \det(\int_{[a,b]} f_i(x)\chi_{I_j}(x)dx)_{i,j=1}^n \\
(2.3.4) \qquad &= \frac{1}{n!} \int \cdots \int_{a<x_1<\cdots<x_n<b} \det(f_i(x_j))_{i,j=1}^n \det(\chi_{I_i}(x_j))_{i,j=1}^n dx_1 \cdots dx_n \\
&= \frac{1}{n!} \int_{I_n} \cdots \int_{I_1} \det(f_i(x_j))_{i,j=1}^n dx_1 \cdots dx_n,
\end{aligned}
$$

where each $\chi_{I_j}(x), 1 \leq j \leq n$, is the characteristic function of I_j. Then (3) $\rightarrow$ (1) follows immediately from (2.3.4).

(1) $\rightarrow$ (3). By (1) $\leftrightarrow$ (2) and (2.3.4), if

$$\det(f_i(x_j))_{i,j=1}^n = 0$$

11

on an open subset of $\mathcal{P}$, we may choose $I_1 < \cdots < I_n$ contained in $[a, b]$ whose product is in this open set and thus $(\int_{I_j} f_i(x)dx)_{i,j=1}^n = 0$. This implies that $(1) \rightarrow (3)$.

$(2) \leftrightarrow (4)$. Suppose that $\{u_i\}_{i=1}^n$ is an H_I-system and suppose to the contrary that there exist n points $\{s_i\}_{i=1}^n$ such that $a < s_1 < \cdots < s_n < b$ and $\{s_i\}_{i=1}^n \cap V = \emptyset$ and there exists a function $g \in [f_1, \ldots, f_n] - \{0\}$ satisfying $g(s_i) = 0$, $i = 1, \ldots, n$. Since $Z(g)$ is nowhere dense in $[a, b]$, from Theorem 2.45 in Schumaker[20], it follows that $g(t) = 0$ for all $t \in [a, s_1] \cup [s_n, b]$, which contradicts (2). Conversely, we easily see that (2) holds under the assumption of (4).

$(2) \leftrightarrow (5)$. From Proposition 2.2.3, the equivalence of (2) and (5) immediately follows.

Remark. (1) Here we show the diagrams of relations between approximating spaces of $C[a, b]$.

$$H_{\mathcal{P}} - spaces \overset{(1)}{\leftrightarrow} C - spaces \overset{(2)}{\leftrightarrow} T_{\mathcal{P}} - spaces \overset{(3)}{\rightarrow} WT_{\mathcal{P}} - spaces$$
$$\downarrow (4) \qquad\qquad \downarrow (5) \qquad\qquad \nearrow (7)$$
$$AC - spaces \underset{(6)}{\leftarrow} H_I - spaces$$

From Proposition 2.2.1, (1) follows, but classically (1) was proven by the results in Young[23] and Haar[8]. Since every $H_{\mathcal{P}}$-space is a $T_{\mathcal{P}}$-space in $C[a, b]$, (2) follows. (3) and (4) are evident. By the above proposition,(5), (6), (7) hold. In particular, by Proposition 2.3.1-(2), every H_I-space is an AC-space.

(2) Basic properties of H_I-systems are studied in Kitahara[14]

Let G be an n-dimensional subspace of $C[a, b]$. For the subspace G, we set the following three subsets of $C[a, b]$:

$A_G^{(1)} = \{f \mid$ the error function $e = f - g$ has $n + 1$ alternating extremal points in $[a, b]$ for any $g \in P_G(f)$, i.e., there exist $n + 1$ points $(a \leq)x_1 < \cdots < x_{n+1}(\leq b)$ such that $|e(x_i)| = \|e\|$, $i = 1, \ldots, n + 1$, and $e(x_i)e(x_{i+1}) \leq 0$, $i = 1, \ldots, n \}$

$A_G^{(2)} = \{f \mid$ the error function $e = f - g$ has $n + 1$ alternating extremal points in $[a, b]$ for some $g \in P_G(f) \}$

$SU_G = \{f \mid f \in C[a, b]$ possesses a strongly unique best approximation from G,i.e., there exist a $g \in G$ and a positive number r, depending only on f, such that $\|f - h\| \geq \|f - g\| + r\|g - h\|$ for all $h \in G \}$.

Now we show some characterizations of $T_{\mathcal{P}}$-spaces, $WT_{\mathcal{P}}$-spaces and H_I-spaces by inclusion relations among $C[a, b], U_G, SU_G, A_G^{(1)}$ and $A_G^{(2)}$.

Theorem 2.3.2. *Let G be an n-dimensional subspace of $C[a,b]$. The following statements are equivalent:*

(1) *G is a T_P-space.*

(2) $A_G^{(1)} = C[a,b]$.

(3) $U_G = SU_G$.

(4) $A_G^{(1)} = U_G \cup L$, *where L denotes the set of all real-valued linear functions on* $[a,b]$.

(1) $\leftrightarrow$ (2) is due to Theorem 8 in Handscomb, Mayers and Powell[9]. (1) $\leftrightarrow$ (3) is proven by McLaughlin and Somers [18;Theorem]. Kitahara [13;Theorem] shows (1) $\leftrightarrow$ (4).

Theorem 2.3.3(Jones and Karlovitz[11;Theorem and Corollary])

Let G be an n-dimensional subspace of $C[a,b]$. The following statements are equivalent:

(1) *G is a WT_P-space.*

(2) $A_G^{(2)} = C[a,b]$.

(3) $U_G \subset A_G^{(2)}$.

Theorem 2.3.4. *Let G be an n-dimensional subspace of $C[a,b]$. The following statements are equivalent:*

(1) *G is an H_I-space.*

(2) $A_G^{(1)}$ *is dense in* $C[a,b]$.

(3) $U_G \subset A_G^{(2)}$ *and U_G is dense in* $C[a,b]$.

Proof. Since the equivalence of (1) and (2) can be extended to $C_0(Q)$, we prove this in 2.4 (see Theorem 2.4.3). (1) $\to$ (3) follows immediately from Proposition 2.3.1. By the equivalence of (1) and (2), (3) $\to$ (1) is easily shown.

2.4 Characterizations of Approximating Spaces of $C_0(Q)$

The diagrams of relations between approximating spaces of $C_0(Q)$ are as follows:

$$H_P - spaces \ \leftrightarrow \ C - spaces \ \leftarrow \ T_P - spaces \ \to \ WT_P - spaces$$
$$\downarrow$$
$$AC - spaces$$

13

First, we review some results which are extensions of Theorem 2.3.2 and Theorem 2.3.3.

For an n-dimensional subspace G of $C_0(Q)$, $A_G^{(1)}$ and $A_G^{(2)}$ denote subsets of $C_0(Q)$ which are defined in the same manner as in the section 2.3.

Theorem 2.4.1.(Deutsch, Nürenberger and Singer[5;Theorem 5.1]) *Let Q be a locally compact subset of R which contains at least $(n + 1)$ points. Let G be an n-dimensional subspace of $C_0(Q)$. The following statements are equivalent:*

(1) *G is a C and WT_P-space $(= T_P$-space).*

(2) *$A_G^{(1)} = C_0(Q)$.*

The following theorem is a part of Theorem 4.1 in Deutsch, Nürenberger and Singer[5].

Theorem 2.4.2. *Let Q be a locally compact subset of R which contains at least $(n+1)$ points. Let G be an n-dimensional subspace of $C_0(Q)$. The following statements are equivalent:*

(1) *G is a WT_P-space.*

(2) *$A_G^{(2)} = C_0(Q)$.*

Now we show an extension of Theorem 2.3.4 to $C_0(Q)$.

Theorem 2.4.3. *Let Q be a locally compact subset of R which contains at least $(n+1)$ points. Let G be an n-dimensional subspace of $C_0(Q)$. The following statements are equivalent:*

(1) *G is an AC and WT_P-space.*

(2) *$A_G^{(1)}$ is dense in $C_0(Q)$.*

To prove this, we prepare several lemmas which are independently interesting. (Simpler proofs may be given.),

Lemma 2.4.4. *Let Q be a locally compact subset of R which contains at least $(n + 1)$ points. Let G be an n-dimensional subspace of $C_0(Q)$. If $A_G^{(1)}$ is dense in $C_0(Q)$, G is a WT_P-space.*

Proof. By Theorem 2.4.2, we only have to show that, for each $f \in C_0(Q)$, there exists a $g \in P_G(f)$ such that $f - g$ has at least $n + 1$ alternating extremal points.

For an arbitrary $f \in C_0(Q) - G$, let $d = d(f, G) := \inf_{h \in G} \|f - h\| > 0$. Since $A_G^{(1)}$

14

is dense in $C_0(Q)$, we can take a sequence $\{f_n\}$ of $A_G^{(1)}$ such that $\lim_n f_n = f$ and $d(f_n, G) > d/2$ for each $n \in N$. For each $n \in N$, let g_n denote a function of $P_G(f_n)$. Since $\{g_n\}$ contains a converging subsequence, without loss of generality, we assume that $\{g_n\}$ converges to g. Noting that $d(f, G) = \lim_n d(f_n, G) = \|f - g\|$, g belongs to $P_G(f)$. Now we consider the function such that

$$k(x) = \max_{n \in N \cup \{0\}} \{|f_n(x) - g_n(x)|\}, \quad x \in Q,$$

where $f_0 = f$, $g_0 = g$. Since the sequence $\{|f_n(x) - g_n(x)|\}_n \cup \{|f(x) - g(x)|\}$ is compact in $C_0(Q)$, it is equicontinuous by Ascoli Theorem (see p.233 in Kelly[12]). Hence, $k(x)$ is continuous on Q. Moreover, for an arbitrary $\epsilon > 0$, we have

$$S_\epsilon(k) = \{x \mid x \in Q, |k(x)| \geq \epsilon\} = \bigcup_{n=0}^{\infty} S_\epsilon(|f_n - g_n|)$$

$$\subset \bigcup_{n=1}^{p} S_\epsilon(|f_n - g_n|) \cup S_{\epsilon/2}(|f - g|),$$

where p is a positive integer satisfying $\|(f_m - g_m) - (f - g)\| < \epsilon/2$ for all $m \geq p$. Hence, $k(x)$ belongs to $C_0(Q)$. Let C_i be a set of $n + 1$ alternating extremal points $x_1^{(i)} < x_2^{(i)} < \cdots < x_{n+1}^{(i)}$ of $f_i - g_i$ for each $i \in N$. Since $S_{d/2}(k)$ is compact and contains C_i for all $i \in N$, without loss of generality, we can assume each sequence $\{x_j^{(i)}\}$, $j = 1, 2, \ldots, n + 1$, converges to $x_j \in Q$. Since $\{f_n - g_n\}$ converges to $f - g$, we easily see that $x_1 < x_2 < \cdots < x_{n+1}$ and that $\{x_i\}_{i=1}^{n+1}$ is a set of $n + 1$ alternating extremal points of $f - g$. This completes the proof.

We restate this in Proposition 2.2.3.

Lemma 2.4.5. *Let Q be a locally compact subset of R which contains at least n points. Let G be an n-dimensional subspace of $C_0(Q)$. G is an AC-space if and only if, on each open subset $U \subset Q$, at most $n - N_n(U)$ linearly independent functions in G can vanish identically.*

Lemma 2.4.6. (p.84 in Feinerman and Newman[6]) *Let $s(x)$ be a sign function with $s(0) = 0$. If M is a proper subspace of R^n, then there exists a vector $(\sigma_1, \ldots, \sigma_n)$ of R^n such that each $\sigma_i, 1 \leq i \leq n$, is 1 or -1 and $(\sigma_1, \ldots, \sigma_n) \neq (s(x_1), \ldots, s(x_n))$ for all $(x_1, \ldots, x_n) \in M$.*

Lemma 2.4.7. (Blatter, Morris and Wulbert[3;p. 13, Theorem]) *Let T be a compact space and let G be a finite dimensional subspace of $C(T)$. Then the following statements are equivalent:*

15

(1) *For any* $f \in C(T)$, *for each* $g \in P_G(f)$ *and for any sequence* $\{f_n\} \subset C(T)$ *which converges to* f, $\lim_{n \to \infty} d(g, P_G(f_n)) = 0$.

(2) *For any* $f \in C(T)$ *such that* $o \in P_G(f)$, $\cap \{Z(g) \mid g \in P_G(f)\}$ *is open in* T, *where* $Z(g) = \{x \mid x \in T, g(x) = 0\}$.

Using an analogous method to the proof of Theorem 1.4 in Sommer[21], we have

Lemma 2.4.8. *Let* $\dot{Q}$ *be an infinite locally compact subset of* R *and let* $Q \cap (-\infty, a]$, $Q \cap [a, b]$, $Q \cap [b, +\infty)$ *be infinite subsets of* Q *for* a *and* b, *where* $a < b$. *If* G *is a finite dimensional* WT_P-*space of* $C_0(Q)$, $G|_{Q'}$ *obtained by restricting* G *to* $Q' = Q \cap [a, b]$ *is a* WT_P-*space of* $C_0(Q')$.

Lemma 2.4.9. *Let* X *be an infinite subset of* R *and let* G *be an* n-*dimensional* WT_P-*space of* $F(X)$ *spanned by* $\varphi_1, \ldots, \varphi_n$. *If no* f *in* $G - \{o\}$ *vanishes on a nonempty subset of the forms* $X \cap [a, b]$, $X \cap (-\infty, c]$ *and* $X \cap [d, +\infty)$ *then there exist* $n + 1$ *points* $x_1 < \cdots < x_{n+1}$ *of* X *satisfying*

$$D \begin{pmatrix} \varphi_1 & \cdots & \varphi_n \\ x_{k_1} & \cdots & x_{k_n} \end{pmatrix} := \det \begin{pmatrix} \varphi_1(x_{k_1}) & \cdots & \varphi_1(x_{k_n}) \\ \vdots & \cdots & \vdots \\ \varphi_n(x_{k_1}) & \cdots & \varphi_n(x_{k_n}) \end{pmatrix} \neq 0$$

for any n *distinct points of* $\{x_i\}_{i=1}^{n+1}$.

Proof. We take $n + 1$ points $x_1 < x_2 < \cdots < x_{n+1}$ of X satisfying the following conditions:

(i) $x_i \notin \{x \mid x \in X, \varphi_j(x) = 0, j = 1, \ldots, n\}$, $i = 1, 2, \ldots, n + 1$

(ii) the subsets of the form $X \cap (-\infty, x_1]$, $X \cap [x_1, x_2]$, ..., $X \cap [x_n, x_{n+1}]$, $X \cap [x_{n+1}, +\infty)$ are nonempty.

Suppose that $D \begin{pmatrix} \varphi_1 & \cdots & \varphi_n \\ x_{k_1} & \cdots & x_{k_n} \end{pmatrix} = 0$ for some n distinct points $x_{k_1}, \ldots, x_{k_n}$ of $\{x_i\}_{i=1}^{n+1}$. Then, there is an $f \in G - \{o\}$ with $f(x_{k_i}) = 0$, $i = 1, \ldots, n$. By Lemma 2.4.12 (= Theorem 1 in Stockenberg[22]), f vanishes on at least one subset of $X \cap (-\infty, x_1]$, $X \cap [x_1, x_2]$, ..., $X \cap [x_n, x_{n+1}]$, $X \cap [x_{n+1}, +\infty)$. This contradicts the hypothesis. Hence, $\{x_i\}_{i=1}^{n+1}$ are points which we require.

Lemma 2.4.10. *Let* Q *be a locally compact subset of* R *and let* G *be a finite dimensional subspace of* $C_0(Q)$. *Suppose that* f *in* $C_0(Q)$ *has a unique best approximation* g *from* G. *Then, for any* $\epsilon > 0$, *there exists a* $\delta > 0$ *such that* $d(g, P_G(h)) < \epsilon$ *for all* h *in* $C_0(Q)$ *with* $\|f - h\| < \delta$.

16

Now we can show

Proof of Theorem 2.4.3. (Necessity) Since $(C_0(Q), \|\cdot\|)$ is a Banach space and G is an AC-space, $\{f \mid f \in C_0(Q), P_G(f) \text{ is not a singleton }\}$ is a set of the first category. Hence, $U_G = \{f \mid f \in C_0(Q), P_G(f) \text{ is a singleton }\}$ is dense in $C_0(Q)$. Since G has WT_P-property, $A_G^{(1)}$ contains U_G, which leads to the necessity.

(Sufficiency) By Lemma 2.4.4, G is a WT_P-space. Suppose that G is not an AC-space which is a WT_P-space of dimension n. By Lemma 2.4.5, there is an open subset U of Q on which more than $n - N_n(U) + 1$ linearly independent functions in G can vanish identically. Now we will prove that $A_G^{(1)}$ is not dense in $C_0(Q)$.

Let U be a finite set $\{x_1, \ldots, x_k\}$, $1 \le k \le n$. Since $\dim G|_U \le k - 1 = q$, let $v_1, \ldots, v_q$ be functions of G which form a base of $G|_U$. Set

$$(2.4.1) \quad L = \sup\{\| \textstyle\sum_{i=1}^{q} a_i v_i \| \mid \max_{1 \le j \le k} | \textstyle\sum_{i=1}^{q} a_i v_i(x_j)| \le 1\} \ (< +\infty).$$

By Lemma 2.4.6, there exists a vector $(\sigma_1, \ldots, \sigma_k)$ of R^k such that each $\sigma_i = 1$ or -1, $1 \le i \le k$, and $(\sigma_1, \ldots, \sigma_k) \ne (s(v(x_1)), \ldots, s(v(x_k)))$ for all $v \in G|_U$. Let us consider a function of $C_0(Q)$ such that

$$(2.4.2) \qquad \begin{aligned} f(x_i) &= \sigma_i M & i &= 1, \ldots, k, \\ f(x) &= 0 & x &\in Q - \{x_1, \ldots, x_k\}, \end{aligned}$$

where M is a positive number much larger than L. Then, we can easily see $o \in P_G(f)$. For a sufficiently small $\epsilon > 0$ and any g in $C_0(Q)$ with $\|f - g\| < \epsilon$, by Lemma 2.4.7, there exists a best approximation $h = \sum_{i=1}^{q} a_i v_i$ to g on U such that $\max_{1 \le j \le k} |h(x_j)| \le 1$. By (2.4.1) and (2.4.2), h is a best approximation to g on Q from G and $g - h$ does not have $n + 1$ alternating extremal points. Since g is arbitrary, $A_G^{(1)}$ is not dense in $C_0(Q)$.

When U is a finite subset of Q consisting of more than $n + 1$ points, we can reduce to the previous case.

Let U be an infinite subset of Q. Without loss of generality, we can assume that

(i) U has no isolated points,

(ii) U contains an infinite subset of Q of the form $F = Q \cap [a, b]$.

If a nonzero function in $G|_F$ vanishes on a subset $Q \cap [c, d]$, $a < c < d < b$, then we retake $Q \cap [c, d]$ as F. By repeating this procedure at most $n - 1$ times, we reach one of the following two situations:

$(iii)_1$ all functions in G vanish identically on $U = Q \cap [\alpha, \beta]$,

$(iii)_2$ each nonzero function in $G|_F$ does not vanish identically on any open subset of F,i.e., $G|_F$ is an AC-space.

Under (i), (ii) and $(iii)_1$, we clearly see that $A_G^{(1)}$ is not dense in $C_0(Q)$. Hence, we show nondensity of $A_G^{(1)}$ in $C_0(Q)$ under (i), (ii) and $(iii)_2$. Let $p = \dim G|_F (< n)$ and let $v_1, \ldots, v_p$ be functions of G which form a base of $G|_F$. By Lemma 2.4.8 and 2.4.9, $G|_F$ is a WT_P-space and we can find $p + 1$ interior points $\{y_i\}_{i=1}^{p+1}$, $y_1 < \cdots < y_{p+1}$, in F such that

$$(2.4.3) \qquad D \begin{pmatrix} v_1 & \cdots & v_p \\ y_{i_1} & \cdots & y_{i_p} \end{pmatrix} \neq 0$$

for any p distinct points of $\{y_i\}_{i=1}^{p+1}$. Set

$$S = \sup \{ \| \sum_{i=1}^{p} a_i v_i \| \mid \sup_{x \in F} | \sum_{i=1}^{p} a_i v_i(x) | \leq 1 \} (< +\infty).$$

Let $U_i = F \cap (a_i, b_i)$ be an open neighbourhood of y_i in Q, $1 \leq i \leq p + 1$, such that the closure of U_i, $1 \leq i \leq p + 1$, is compact and $U_j \cap U_k = \emptyset$ for $j \neq k$. By Urysohn's lemma (see p.75 in Hewitt and Stromberg[10]), there are functions $f_i, 1 \leq i \leq p + 1$, in $C_0(Q)$ satisfying that

$$f_i(y_i) = 1,$$

$$0 < f_i(y) < 1 \qquad y \in U_i - \{y_i\},$$

$$f_i(y) = 0 \qquad y \in Q - U_i.$$

Let us consider a function g of $C_0(Q)$ such that

$$(2.4.4) \qquad g(x) = \sum_{i=1}^{p+1} (-1)^i f_i(x) \cdot C,$$

where C is a positive number much larger than S. By $(2.4.3)$, $(2.4.4)$ and WT_P-property of $G|_F$, $P_{G|_F}(g) = \{o\}$ and $o \in P_G(g)$. From Lemma 2.4.10, we see that, for sufficiently small $\delta > 0$ and any f in $C_0(Q)$ with $\|f - g\| < \delta$, $\sup_{x \in F} |h(x)| \leq 1$ for all $h \in P_{G|_F}(f)$. Using $(2.4.3)$ and $(2.4.4)$, we observe that each h of $P_{G|_F}(f)$ belongs to $P_G(f)$ and $f - h$ does not have $n + 1$ alternating extremal points in Q. This implies that $A_G^{(1)}$ is not dense in $C_0(Q)$. In any case, we obtain the conclusion.

When Q is a union of disjoint open subintervals (a, b) of R, the following proposition is obtained from a similar proof to Theorem in Kitahara[13].

Proposition 2.4.11. *Let Q be a union of disjoint open subintervals (a, b) of R and let G be an n-dimensional subspace of $C_0(Q)$. The following statements are equivalent:*

(1) *G is a WT_P and H_P-space $(= T_P$-space $)$.*

(2) *$A_G^{(1)} = U_G$ and, for each point $x \in Q$, there exists a nonnegative function $f \in A_G^{(1)}$ such that $f(x) = \|f\|$.*

Before proving this, some preparations are needed.

Definition. Let T be a totally ordered set with an order $<$.

(1) For a function $f \in F(T)$, two zeros x_1, x_2 $(x_1 < x_2)$ of f are said to be *separated* if there is a point $x_0 \in T$ such that $x_1 < x_0 < x_2$ and $f(x_0) \neq 0$.

(2) Let G be a finite dimensional subspace of $F(T)$. A point $x_0 \in T$ is called a *vanishing point with respect to* G if $g(x_0) = 0$ for all $g \in G$. The set of all vanishing points with respect to G is denoted by $V(G)$. If $V(G) = \emptyset$, G is called a *nonvanishing space*.

Lemma 2.4.12. (Stockenberg[22;Theorem 1]) *Let T be a totally ordered set and G be an n-dimensional WT_P-space of $F(T)$. Then the following holds:*

(1) *If there is a $g \in G$ with separated, nonvanishing zeros $x_1 < \ldots < x_n$, then $g(x) = 0$ for all x with $x \leq x_1$ or $x \geq x_n$.*

(2) *No $g \in G$ has more than n separated, nonvanishing zeros.*

From Proposition 2.7 and Theorem 3.1 in Deutsch, Nürenberger and Singer[5], we get

Lemma 2.4.13. *Let Q be a locally compact subset of R and let G be an n-dimensional space of $C_0(Q)$ which does not contain any delta functions, i.e., a characteristic function of a point in Q. Then $A_G^{(1)} \supset U_G$ if and only if G is a WT_P-space.*

Now we are in position to state

Proof of Proposition 2.4.11. $(1) \to (2)$ is trivial. Hence, it is sufficient to verify that G is a WT_P and H_P-space under the assumption (2).

$(2) \to (1)$. First we show that G is a nonvanishing WT_P-space. Since G does not contain a delta function, by Lemma 2.4.13, G is a WT_P-space. Suppose that there is a vanishing point x_0 with respect to G. By the hypothesis, there exists a nonnegative function $f \in A_G^{(1)}$ such that $f(x_0) = \|f\|$. Therefore, one of the best approximations to f from G is 0. But, this contradicts the condition of $A_G^{(1)}$. Hence, when $n = 1$, G

19

is a WT_P and H_P-space. In the rest of the proof, we assume $n \geq 2$.

Next, we show that any function $g \in G - \{0\}$ does not vanish identically on any nonempty subset $Q \cap [a, b]$, where $[a, b]$ is a nondegenerate subinterval. Suppose that a $g \in G - \{0\}$ vanishes on a nonempty subset $Q' = Q \cap [c, d]$, $c < d$ (For convenience, we call this property of G $v.i.$-property.). Then, from the fact that $V(G) = \emptyset$ and Lemma 2.4.8, $G_1 = G|_{Q'}$ is a nonvanishing WT_P-space such that $\dim G|_{Q'} < n$. When G_1 has $v.i.$-property, we can consider a nonvanishing WT_P-space $G_1|_{Q \cap [\alpha, \beta]}$ obtained by the same way with respect to G_1, where $c \leq \alpha < \beta \leq d$ and $\dim G_1|_{Q \cap [\alpha, \beta]} < \dim G_1$. By continuing the above procedure at most $n - 1$ times, we consequently obtain a nonvanishing WT_P-space $G|_{Q \cap [\gamma, \delta]}$ without $v.i.$-property, where $Q'' = Q \cap [\gamma, \delta]$, $c \leq \gamma < \delta \leq d$, is a nondegenerate compact interval and $1 \leq m = \dim G|_{Q''} < n$. Now we consider a function $f_0 \in C_0(Q)$, which satisfies the following conditions:

(i) $f_0(x) = 0$ for all $x \in Q - Q''$.

(ii) There are $2(n + m + 2)$ points $\gamma < z_1 < \cdots < z_{2(n+m+2)} < \delta$ of Q'' such that $|f_0(z_i)| = \|f_0\| > 0$, $i = 1, 2, \ldots, 2(n + m + 2)$ and $f_0(z_i) \cdot f_0(z_{i+1}) < 0$ for $i = 1, \ldots, 2n + 2m + 3$.

Since G is assumed to have $v.i.$-property, there is a function $h^* \in G - \{0\}$ such that $\|h^*\| < \|f_0\|$ and $h^*(x) = 0$ for $x \in Q \cap [c, d]$. Thus, each function $\lambda \cdot h^*$, $0 \leq \lambda \leq 1$, is a best approximation to f_0 from G because G is a WT_P-space and the error function $f_0 - \lambda \cdot h^*$ has an alternating set of $(n + 1)$-points in Q. On the other hand, since $G|_{Q''}$ is a nonvanishing WT_P-space without $v.i.$-property, Lemma 2.4.12, each function $f \in G|_{Q''} - \{0\}$ has at most m zeros in Q''. Provided that there exists a best approximation to f_0 from G which has an alternating set of at most n-points in Q'', then it has at least $(m + 1)$ zeros in Q''. This leads to a contradiction. Eventually, by these facts, we conclude that f_0 is contained in $A_G^{(1)}$ but not U_G, which is contrary to the assumption. Hence, G is a nonvanishing WT_P-space without $v.i.$-property, that is, a WT_P and H_P-space.

Corollary 2.4.14. *Let Q be a union of disjoint open subintervals (a, b) of R and let G be an n-dimensional nonvanishing WT_P-space of $C_0(Q)$. Then G has $v.i.$-property if and only if $U_G \subsetneq A_G^{(1)}$.*

Proof. First suppose that G has $v.i.$-property. By using the proof of Proposition 2.4.11, we easily observe that $U_G \subsetneq A_G^{(1)}$.

Next, suppose that any nonzero function in G has at most n zeros, because, by

Lemma 2.4.12, this is equivalent to the fact that G does not have $v.i.$-property. For any function $f \in A_G^{(1)}$, let g_1, g_2 be best approximations to f from G. Since the function $(g_1 + g_2)/2$ is also a best approximation to f, $f - (g_1 + g_2)/2$ has $(n+1)$ alternating extremal points $\{x_i\}_{i=1}^{n+1}$ in Q. Hence, we have

$$\|f - (g_1 + g_2)/2\| = |f(x_i) - (g_1(x_i) + g_2(x_i))/2|$$
$$\leq (1/2) \cdot \{|f(x_i) - g_1(x_i)| + |f(x_i) - g_2(x_i)|\}$$
$$i = 1, 2, \ldots, n+1,$$

which means that

$$f(x_i) - g_1(x_i) = f(x_i) - g_2(x_i) \quad i = 1, 2, \ldots, n+1.$$

Thus, $g_1 - g_2$ has at least $(n+1)$ zeros, which leads to the fact that g_1 is identical with g_2 on Q. Hence, we obtain $A_G^{(1)} = U_G$.

Corollary 2.4.15. *Let Q be a union of disjoint open subintervals (a, b) of R and let G be an n-dimensional space of $C_0(Q)$. If $A_G^{(1)} = U_G$, then G is a WT_P and AC-space.*

Proof. By Lemma 2.4.13, it is clear that G is a WT_P-space. For any $f \in G - \{0\}$, $Z(f)$ is nowhere dense in Q, because $A_G^{(1)} = U_G$ and the proof of Proposition 2.4.11. Hence, G is an AC-space by Proposition 2.2.3.

Remark. (1) The assertion in Corollary 2.4.14 does not always hold for finite vanishing points instead of no vanishing points with respect to the space G. For example, on $C[0, \pi]$, let $G = \{\lambda \cdot \sin x \mid \lambda \in R\}$. Clearly, G is a one dimensional WT_P-space which does not have $v.i.$-property but 2 vanishing points in $[0, \pi]$. Thus, the best approximation to the linear function $f(x) = -2x + \pi$ is not unique and any best approximation to it has an alternating set of 2 points, 0 and π. Therefore, we obtain $A_G^{(1)} \underset{\neq}{\supset} U_G$.

(2) By generalizing the example of (1), it can be shown that $U_G \underset{\neq}{\subsetneq} A_G^{(1)}$ for every n-dimensional WT_P-space G of continuously differentiable functions such that G has at least $(n+1)$ vanishing points.

(3) From the example of (1), we observe that the converse of Corollary 2.4.15 does not always hold.

2.5 A Characterization of H_P-Spaces of $C(R)$

Let $C(R)$ be the space of all real-valued continuous functions on the real line R. One can easily see that if a system $\{u_i\}_{i=1}^n$ in $C(R)$ satisfies one of the following equivalent conditions ((a) is the definition of H_P-systems.), it is an H_P-system.

(a) For any $\{x_i\}_{i=1}^n \subset R$ such that $x_i \neq x_j$ for $i \neq j$, the n-th determinant

$$\det \begin{pmatrix} u_1(x_1) & \cdots & u_n(x_1) \\ \vdots & \cdots & \vdots \\ u_1(x_n) & \cdots & u_n(x_n) \end{pmatrix} \neq 0.$$

(b) For any n-points (x_i, y_i), $i = 1, \ldots, n$, in R^2 satisfying $x_i \neq x_j$ for $i \neq j$, there exists a unique $u \in [u_1, \ldots, u_n]$ such that $u(x_i) = y_i$, $i = 1, \ldots, n$.

(b)' For any $n + 1$ decomposition $\{A_i\}_{i=1}^{n+1}$ of R^2 such that $A_1 = \{(x_1, y_1)\}$, ..., $A_n = \{(x_n, y_n)\}$, $A_{n+1} = R^2 - \cup_{i=1}^n A_i$ and $x_i \neq x_j$ for $i \neq j$, there is a unique $u \in [u_1, \ldots, u_n]$ which passes through points in A_i, $i = 1, \ldots, n + 1$.

In (b), we sometimes write $u[(x_1, y_1), \ldots, (x_n, y_n)]$ for the linear combination of $\{u_i\}_{i=1}^n$ such that $u(x_i) = y_i$, $i = 1, \ldots, n$.

The purpose of this section is to give a characterization of H_P-systems $\{u_i\}_{i=1}^n$ ($n \geq 2$) in terms of appropriate $n + 2$ decompositions of R^2 as (b)'. To be more precise, we shall show

Theorem 2.5.1. *Let $\{u_i\}_{i=1}^n$ ($n \geq 2$) be a system in $C(R)$. Then the following statements are equivalent:*

(1) *$\{u_i\}_{i=1}^n$ is an H_P-system on R.*

(2) *For any $n + 2$ decomposition $\{A_i\}_{i=1}^{n+2}$ of R^2 with (*)-property (see Definition 2), there exist a subset $\{A_{i_k}\}_{k=1}^{n+1}$ of $\{A_i\}_{i=1}^{n+2}$ and a $u \in [u_1, \ldots, u_n]$ which passes through points in A_{i_k}, $k = 1, \ldots, n + 1$.*

Before proving this, some definitions and a lemma are prepared.

Definition 1. Let $\{(x_i, y_i)\}_{i \in I}$ be a set of points in R^2, and let $\{A_i\}_{i=1}^n$ be a family of n subsets of R^2 such that $A_i \cap A_j = \emptyset$ for $i \neq j$.

(1) The points of (x_i, y_i), $i \in I$, are said to be *distinct* if $x_i \neq x_j$ for all $i \neq j$.

(2) If there are k ($k \leq n$) distinct points $a_1, \ldots, a_k$ in R^2 such that $\{a_1, \ldots, a_k\} \subset \cup_{i=1}^n A_i$ and each A_i, $i = 1, \ldots, n$, contains at most one point $\{a_i\}_{i=1}^k$, then $a_1, \ldots, a_k$ are called *distinct with $\{A_i\}_{i=1}^n$*.

22

Definition 2. Let ℓ_α denote the straight line $\{(\alpha, y) \mid y \in R\}$. If an $n+2$ decomposition $\{A_i\}_{i=1}^{n+2}$, $n \in N$, satisfies (1) and (2), then we call it an $n+2$ decomposition with (*)- *property*.

(1) At least one A_j of $\{A_i\}_{i=1}^{n+2}$ has interior points.

(2) If for some A_j of $\{A_i\}_{i=1}^{n+2}$ and $\alpha \in R$, A_j contains ℓ_α, then there are at least $n+1$ distinct points with $\{A_i\}_{i=1}^{n+2}$. Otherwise, there are $n+2$ distinct points with $\{A_i\}_{i=1}^{n+2}$.

Lemma 2.5.2. *Let $\{u_i\}_{i=1}^n$, $n \in N$, be an H_P-system in $C(R)$ and let (x_i, y_i), $i = 1, \ldots, n$, be n distinct fixed points in R^2. For a variable point (x, y) with $x \neq x_i$, $i = 2, \ldots, n$, each $c_j(x, y)$, $j = 1, \ldots, n$, denotes the unique coefficient of u_j such that $\sum_{i=1}^n c_i(x, y) u_i$ passes through (x, y) and (x_i, y_i), $i = 2, \ldots, n$. If (x, y) converges to (x_1, y_1), then $c_j(x, y)$, $j = 1, \ldots, n$, converges to $c_j(x_1, y_1)$.*

Proof. By using Cramer's rule, one can easily show the continuity of $c_j(x, y)$, $j = 1, \ldots, n$, at (x_1, y_1).

Now we are in position to show

Proof of Theorem 2.5.1. $(2) \to (1)$. We prove (b) in the equivalent conditions of H_P-systems. Let (x_i, y_i), $i = 1, \ldots, n$, be any n distinct points in R^2. As an $n+2$ decomposition of R^2 with (*)-property, we consider $A_1 = \{(x_1, y_1)\}$, $\ldots$, $A_n = \{(x_n, y_n)\}$, $A_{n+1} = \{(x_n, y_{n+1})\}(y_n \neq y_{n+1})$, $A_{n+2} = R^2 - \cup_{i=1}^{n+1} A_i$. By the hypothesis, there is a $v_0 \in [u_1, \ldots, u_n]$ which passes through points in A_i, $i = 1, \ldots, n$ and A_{n+2} or through points in A_i, $i = 1, \ldots, n-1$, A_{n+1} and A_{n+2}. In the former case, v_0 is a linear combination of $\{u_i\}_{i=1}^n$, which is required. Let us take the latter case. As another $n+2$ decomposition of R^2 with (*)-property, we prepare $B_1 = \{(x_1, 0)\}$, $\ldots$, $B_{n-1} = \{(x_{n-1}, 0)\}$, $B_n = \{(x_n, r)\}(r \neq 0)$, $B_{n+1} = \{(x_n, s)\}(s \neq 0)$, $B_{n+2} = R^2 - \cup_{i=1}^{n+2} B_i$. From the hypothesis, we get a $v_1 \in [u_1, \ldots, u_n]$ such that $v_1(x_i) = 0$, $i = 1, \ldots, n-1$, and $v_1(x_n) \neq 0$. Hence, for some $\gamma \in R$, $v_0 + \gamma v_1$ passes through points (x_i, y_i), $i = 1, \ldots, n$.

$(1) \to (2)$. We prove this through considering three cases.

(Case 1) Suppose that, for some $\alpha_0 \in R$, ℓ_{α_0} consists of points in at most two $A's$ of $\{A_i\}_{i=1}^{n+2}$. Since $\{A_i\}_{i=1}^{n+2}$ has (*)-property, there are n distinct points (x_{i_k}, y_{i_k}) with $\{A_i\}_{i=1}^{n+2}$ such that $(x_{i_k}, y_{i_k}) \in A_{i_k}$ and $\ell_{\alpha_0} \cap A_{i_k} = \emptyset$, $k = 1, \ldots, n$. By the hypothesis of $\{u_i\}_{i=1}^n$, there exists a $u_0 \in [u_1, \ldots, u_n]$ satisfying $y_{i_k} = u_0(x_{i_k})$, $k = 1, \ldots, n$. Since $(\alpha_0, u_0(\alpha_0))$ belongs to some A_j with $j \neq i_k$, $k = 1, \ldots, n$, u_0 is a linear combination

23

of $\{u_i\}_{i=1}^n$ which is required.

In the rest of this proof, we can make the following assumptions without losing any generality:

(A.1) For any $\alpha \in R$, ℓ_α consists of points in at least three $A's$ of $\{A_i\}_{i=1}^{n+2}$.

(A.2) $(0,0)$ is an interior point of A_1 and a closed neighbourhood $U = \{(x,y) \mid \max\{|x|,|y|\} \le \delta\}$ $(\delta > 0)$ of $(0,0)$ is contained in A_1. Let $V = \{(x,y) \mid |x| \le \delta, y > \delta\}$ and let $W = \{(x,y) \mid |x| \le \delta, y < -\delta\}$.

(Case 2) Suppose that there are 2 distinct points z_1, z_2 with $\{A_i\}_{i=2}^{n+2}$ such that $z_1 \in V$ and $z_2 \in W$. Let w_i, $i = 1,\ldots,n+1$, be $n+1$ distinct points with $\{A_i\}_{i=2}^{n+2}$ which can be taken by (*)-property of $\{A_i\}_{i=1}^{n+2}$. Then, among the points z_1, z_2 and w_i, $i = 1,\ldots,n+1$, we choose n distinct points (x_{i_k}, y_{i_k}) with $\{A_i\}_{i=2}^{n+2}$ such that $(x_{i_{k_0}}, y_{i_{k_0}}) \in V$ and $(x_{i_{k_1}}, y_{i_{k_1}}) \in W$ for some k_0, k_1. It is clear the linear combination $u[(x_{i_1}, y_{i_1}), \ldots, (x_{i_n}, y_{i_n})]$ of $\{u_i\}_{i=1}^n$ satisfies (2) (see Fig. 1).

Fig. 1

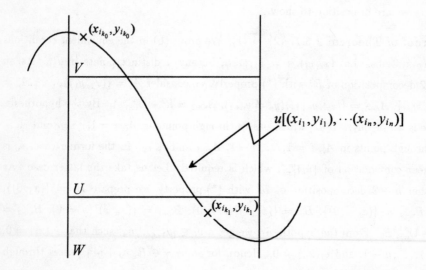

(Case 3) Suppose that A_1 contains V or W, for example, W is contained in A_1. By (A.1), each $\ell_\alpha \cap V$, $|\alpha| \le \delta$, consists of points in at least two $A's$ of $\{A_i\}_{i=2}^{n+2}$. Without loss of generality, we may assume that there is an uncountable set $I \subset (-\delta, \delta)$ and, for each $\alpha \in I$, ℓ_α contains two points (α, y_α), (α, y'_α) satisfying

(2.5.1) $\qquad\qquad (\alpha, y_\alpha) \in A_2, \quad (\alpha, y'_\alpha) \in A_3,$

$$y'_\alpha > y_\alpha.$$

From the uncountability of I and (2.5.1), there are an uncountable subset I' of I and a positive integer n_0 such that, for all $\alpha \in I'$,

(2.5.2) $\qquad\qquad\qquad y'_\alpha \leq n_0$

(2.5.3) $\qquad\qquad\qquad y'_\alpha - y_\alpha \geq 1/n_0.$

By (2.5.2), there exists a monotone sequence $\{\alpha_n\} \subset I'$ convergent to α_0 such that $\{y_{\alpha_n}\}$ and $\{y'_{\alpha_n}\}$ converge to y_{α_0} and y'_{α_0} respectively. Assume that $\{\alpha_n\}$ is monotone increasing. We consider $n-1$ distinct points $\{t_i\}_{i=1}^{n-1}$ with $\{A_i\}_{i=1, i\neq 2,3}^{n+2}$ such that

(2.5.4) $\qquad\qquad\qquad t_1 = (\eta_0, \delta) \in A_1 \quad (-\delta < \eta_0 < \alpha_0)$

and, for $t_i = (r_i, s_i)$, $i = 2, \ldots, n-1$,

(2.5.5) $\qquad\qquad\qquad r_i \in (-\infty, \eta_0) \cup (\alpha_0, +\infty).$

Indeed, these points can be considered by using the (*)-property of $\{A_i\}_{i=1}^{n+2}$. By Lemma 2.5.2, the sequence $\{u[t_1, \ldots, t_{n-1}, (\alpha_m, y'_{\alpha_m})]\}_m$ $(\alpha_m \neq r_i, i = 2, \ldots, n-1)$ of linear combinations of $\{u_i\}_{i=1}^n$ uniformly converges to $u[t_1, \ldots, t_{n-1}, (\alpha_0, y'_{\alpha_0})]$. If n_1 is a sufficiently large positive integer, from (2.5.3), we obtain

(2.5.6) $\qquad\qquad\qquad \eta_0 < \alpha_{n_1-1}$

$$u(\alpha_{n_1-1}) > y_{\alpha_{n_1-1}},$$

where $u = u[t_1, \ldots, t_{n-1}, (\alpha_{n_1}, y'_{\alpha_{n_1}})]$. By (2.5.4) and (2.5.5), (η, δ) $(\eta_0 \leq \eta \leq \alpha_{n_1-1})$, $t_2, \ldots, t_{n-1}$ and $(\alpha_{n_1}, y'_{\alpha_{n_1}})$ are n distinct points with $\{A_i\}_{i=1, i\neq 2}^{n+2}$ and when $\eta = \alpha_{n_1-1}$, $u = u[(\alpha_{n_1-1}, \delta), t_2, \ldots, t_{n-1}, (\alpha_{n_1}, y'_{\alpha_{n_1}})]$ satisfies

(2.5.7) $\qquad\qquad\qquad u(\alpha_{n_1-1}) < y_{\alpha_{n_1-1}}.$

Combining (2.5.6) and (2.5.7) with Lemma 2.5.2, we can find an $\eta_1 (\eta_0 < \eta_1 < \alpha_{n_1-1})$ such that $u[(\eta_1, \delta), t_2, \ldots, t_{n-1}, (\alpha_{n_1}, y'_{\alpha_{n_1}})]$ passes through $(\alpha_{n_1-1}, y_{\alpha_{n_1-1}}) \in A_2$ (see Fig. 2). This linear combination follows condition (2). Hence, in any case we prove that (1) implies (2).

25

Fig. 2

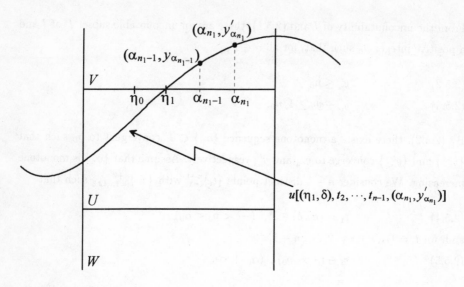

Example. Let us consider an H_P-system $u_1(x) = 1$, $u_2(x) = x$ and let $\{A_i\}_{i=1}^4$ be any 4 decomposition of R^2 satisfying (2) in Definition 2. Without loss of generality, suppose that A_1 is an infinite subset which is not contained in a straight line of R^2. Let y_1 be a straight line which is not a form of $x = \alpha$, $\alpha \in R$, that passes through points of at least two subsets among A_2, A_3 and A_4, for example A_3 and A_4. Indeed, y_1 exists from the hypothesis. When y_1 contains points of three subsets among A_1, A_2, A_3 and A_4, $y_1 = ax + b$ is a straight line which we require. When y_1 is contained in $A_3 \cup A_4$, let us consider a straight line $y_2 = cx + d$ which passes through points of A_1 and A_2 and is not parallel to y_1. Then it is not difficult to verify that y_2 is a straight line which is required.

It seems that Theorem 2.5.1 holds for a family of $n+2$ decomposition of R^2 satisfying weaker conditions than (*)-property, but we are not in position to show any facts about it.

2.6 A Characterization of Spaces Spanned by Infinite Complete T_P-Systems in $C[a, b]$

In this section, we are concerned with the following nonexistence theorem of best approximations by an infinite complete T_P-system $\{u_i\}_{i=1}^{\infty}$, i.e., each system $\{u_i\}_{i=1}^{k}$, $k \in N$, is a T_P-system (see p.76 in Cheney[4]).

Theorem 2.6.1. *Let $\{u_i\}_{i=1}^{\infty}$ be an infinite complete T_P system in $C[a,b]$ and let M be the closed linear subspace of $C[a,b]$ generated by $\{u_i\}_{i=1}^{\infty}$. Then M has the following property:*

$(**)$ *For each function $f \in C[a,b]$ outside M, there is no best approximation $\tilde{f}$ to f from M,i.e., there does not exist $\tilde{f}$ such that $\|f - \tilde{f}\| = \inf_{g \in M} \|f - g\|$.*

First, we make some preparations to show the converse of this theorem.

If $\{u_i\}_{i=1}^{n}$ is a WT_P-system in $C[a,b]$ such that, for any nonzero linear combination v of $\{u_i\}_{i=1}^{n}$, where $Z(v) = \{x | v(x) = 0\}$ is nowhere dense in $[a,b]$, we call $\{u_i\}_{i=1}^{n}$ an H_I-system (see p. 10)

From Theorem 4 in Stockenberg[22], we easily obtain

Lemma 2.6.2. *Let G be a space spanned by an H_I-system $\{u_i\}_{i=1}^{n}$ in $C[a,b]$. Suppose that G contains a strictly positive function and contains two functions r, s such that*

$$(T) \qquad \det \begin{pmatrix} r(a) & r(b) \\ s(a) & s(b) \end{pmatrix} \neq 0$$

Then $\{u_i\}_{i=1}^{n}$ is a T_P-system.

Proof. Suppose that $\{u_i\}_{i=1}^{n}$ is not a T_P-system on $[a,b]$. By the condition (T), there is an $h \in G$ such that $h(a) = h(b) \neq 0$. By Theorem 4-(4) in Stockenberg[22], $f(a) = f(b)$ for any $f \in G$. This contradicts the condition (T).

Now we can state

Theorem 2.6.3. *Let $\{u_i\}_{i=1}^{\infty}$ be an infinite complete H_I-system in $C[a,b]$, i.e., each system $\{u_i\}_{i=1}^{k}$, $k \in N$, is an H_I-system and let M be the closed linear subspace of $C[a,b]$ generated by $\{u_i\}_{i=1}^{\infty}$. Then the following statements are equivalent:*

(1) *There is a positive integer k_0 and each system $\{u_i\}_{i=1}^{k}$ is a T_P-system for $k \geq k_0$.*

(2) *M has $(**)$-property.*

Proof. $(1) \to (2)$. Since the proof can be reduced to that of Theorem 2.6.1, we omit it.

$(2) \to (1)$. First we will prove that some M_{k_1} contains a strictly positive function, where M_k denotes the space spanned by $\{u_i\}_{i=1}^{k}$. Suppose that, for each $n \in N$, any

27

function of M_n has at least one zero on $[a,b]$. Since any function of M also has at least one zero on $[a,b]$, M does not contain 1. As is easily seen, 0, which belongs to M, is a best approximation to 1 from M. This contradicts the fact that M has (**)-property.

Next, we show that some M_{k_2} contains two functions r,s which satisfy the condition (T). Suppose to the contrary that no M_n contains functions that satisfy the condition (T). Since M contains a strictly positive function, M does not contain a continuous function $g(x)$ with $\|g\| = g(a) = -g(b)(\neq 0)$ and 0 is a best approximation to g from M. This contradicts the (**)-property of M. Hence, M_k, $k \geq \max\{k_1, k_2\}$, is a space spanned by an H_I-system that contains two functions r,s satisfying

$$\det \begin{pmatrix} r(a) & r(b) \\ s(a) & s(b) \end{pmatrix} \neq 0.$$

By Lemma 2.6.2, each system $\{u_i\}_{i=1}^{k}$, $k \geq \max\{k_1, k_2\}$, is a T_P-system. This completes the proof.

Remark. If, in Theorem 2.6.3, "infinite complete H_I-system " is replaced with "infinite complete WT_P-system ",i.e., each system $\{u_i\}_{i=1}^{k}$, $k \in N$, is a WT_P-system, we can see that (2) does not always imply (1) by the following example. Let us consider an infinite system $\{u_i\}_{i=0}^{\infty}$ in $C[0,2]$ such that

$$u_0(x) = 1$$

$$u_{2i-1}(x) = \begin{cases} (x-1)^i & x \in [0,1] \\ 0 & x \in [1,2], \quad i = 1,2,\ldots \end{cases}$$

$$u_{2i}(x) = \begin{cases} 1 & x \in [0,1] \\ x^{i^2} & x \in [1,2], \quad i = 1,2,\ldots \end{cases} .$$

By Theorem 4 in Bartelt[2], $\{u_i\}_{i=0}^{\infty}$ is an infinite WT_P-system. Since $Z(u_{2i-1}) = [1,2]$, $i = 1,2,\ldots$, $\{u_i\}_{i=0}^{\infty}$ is not an infinite H_I-system. By Müntz's Theorem (see p.197 in Cheney[4]), $\{1\} \cup \{(x-1)^i\}$ is fundamental in $C[0,1]$ and $\{1\} \cup \{x^{i^2}\}$ is not fundamental in $C[1,2]$. Since each $u_{2i-1}(x)$, $i = 1,2,\ldots$, vanishes on $[1,2]$, the closed subspace M generated by $\{u_i\}_{i=0}^{\infty}$ of $C[0,2]$ consists of real-valued continuous functions whose restrictions to $[1,2]$ belong to the closed subspace M_1 generated by $\{1\} \cup \{x^{i^2}\}$. From this fact, for any continuous function f outside M, we have

$$\inf_{g \in M} \sup_{x \in [0,2]} |f(x) - g(x)| = \inf_{h \in M_1} \sup_{x \in [1,2]} |f(x) - h(x)|.$$

28

Since, in $C[1,2]$, f does not have a best approximation $\tilde{f}$ from M_1 by Theorem 2.6.1, there does not exist a best approximation to f from M in $C[0,2]$. Hence, M has the (**)-property.

2.7 Problems

1. In 2.3, characterizations of Haar-like spaces G are shown by using inclusion relations among $C[a,b]$, U_G, SU_G, $A_G^{(1)}$ and $A_G^{(2)}$. Can we obtain characterizations of Haar-like spaces by other inclusion relations among the sets mentioned above ?

2. In 2.5, a characterization of H_P-spaces of $C(R)$ is shown in terms of decompositions of R^2 with (*)-property. Does Theorem 2.5.1 hold under decompositions of R^2 with a weaker condition than (*)-property ? For example, we make the following conjecture in search of weaker conditions:

Let $\{u_i\}_{i=1}^n$ ($n \geq 2$) be a system in $C(R)$. Then $\{u_i\}_{i=1}^n$ is an H_P-system on R if and only if, for any $(n+2)$ decomposition $\{A_i\}_{i=1}^{n+2}$ of R^2 satisfying (2) in Definition 2,there exist a subset $\{A_{i_k}\}_{k=1}^{n+1}$ of $\{A_i\}_{i=1}^{n+2}$ and a $u \in [u_1,\ldots,u_n]$ which passes through points in A_{i_k}, $k = 1,\ldots,n+1$.

3. Let $\{u_i\}_{i=1}^n$ ($n \geq 2$) be a system in $C(R)$ and, let $\{A_i\}_{i=1}^{n+1}$ of R^2 be a $(n+1)$ decomposition of R^2 satisfing the conditions that
  (i) at least one A_j of $\{A_i\}_{i=1}^{n+1}$ has interior points,
  (ii) there are $(n+1)$ points distinct with $\{A_i\}_{i=1}^{n+1}$.
Then are the following statements equivalent ?
  (1) $\{u_i\}_{i=1}^n$ is an H_P-system on R.
  (2) For any $(n+1)$ decomposition $\{A_i\}_{i=1}^{n+1}$ of R^2 with (i) and (ii), there exist a $u \in [u_1,\ldots,u_n]$ which passes through points in $A_i, i = 1,\ldots,n+1$.

(2) $\to$ (1) is easily obtained but we can't prove whether (2) $\to$ (1) or not.

4. In 2.6, under the assumption that M is a closed subspace of $C[a,b]$ generated by an infinite complete H_I-system, we characterize M with (**)-property. Let M be a closed subspace of $C[a,b]$ generated by an infinite complete WT_P-system. Then can we find a subclass of infinite complete WT_P-systems such that M has (**)-property ?

Chapter 3 Some Topics of Haar-like Spaces of $F[a, b]$

3.1 Introduction.

In this chapter, we consider some topics of T_P, WT_P and H_I-spaces of $F[a, b]$.

In 3.2, the results of Zwick[43] are used to investigate derivative spaces of Haar-like spaces. Zwick[43] gave a characterization of derivative spaces for complete WT_P-spaces of $C^1[a, b]$, and Kitahara[7] showed a corresponding characterization of derivative spaces for complete T_P-spaces of $C^1[a, b]$. In this section, we show characterizations of derivative spaces for complete WT_P-spaces of $AC[a, b]$ or complete T_P-spaces of $AC[a, b]$. These results are extensions of Zwick's and Kitahara's results.

In 3.3, we are concerned with representations of WT_P-systems. Integral representations of WT_P-systems are deeply studied by Rutman[23], Schwenker[25], Zalik[35, 36, 37, 38], Zielke[42] and so on. Here separated representations (see Remark 1 in 3.2) are treated. We get two main results. One is a representation of WT_P-systems $\{u_i\}_{i=1}^n$ in $F[a, b]$ such that $\cap_{i=1}^n Z(u_i)$ is nowhere dense in $[a, b]$. The other is a representation of WT_P-systems in $C^m[a, b]$ (= the space of all real-valued m times continuously differentiable functions on $[a, b]$).

In 3.4, adjoined functions to a given space with a Haar-like condition are considered. Zielke[40,41] and Zalik[33] already showed the existence of adjoined functions to complete T_P-spaces of $F[a, b]$ or adjoined continuous functions to T_P-spaces of $C[a, b]$. In this section, making use of the results in 3.2 and 3.3, we show the existence of continuously differentiable functions which are adjoined to complete T_P-spaces of $C^1[a, b]$ and the existence of adjoint functions to other Haar-like spaces.

In 3.5, the uniqueness of best L^1-approximations in $C[a, b]$ is examined. Recently best L^1-approxiamtions to continuous functions from WT_P-spaces have been studied in Micchelli[17], Sommer[26, 27], Strauss[31] and others, and properties of unicity subspaces in weighted L^1-norms, which are called A-spaces, have been profoundly investigated (e.g. Kroó[12,13,14], Li[16], Pinkus[20], Schmidt[24] and Sommer[28,29,30] and others). In this section, we consider H_I-spaces, which are not A-spaces in general, and study best L^1-approximations to continuous functions from an H_I-space of $C[a, b]$. As a main result, we show a necessary and sufficient condition for any continuous function to have a unique best L^1-approximation from an H_I-space.

In 3.6, we give some problems related to these topics.

3.2 Spaces of Derivatives of Complete WT_I or Complete T_P-Spaces

We begin with definitions and notations.

Definition 1.(see p. 7 and 33 in Zielke[41]) (1) A real-valued function f on a linearly ordered set T is said to have an *alternation of length n*, if there are n points $x_1 < \cdots < x_n$ in T such that $(-1)^i f(x_i)$, is positive for $i = 1, \ldots, n$ or negative for $i = 1, \ldots, n$.

(2) A real-valued function f on $[a, b]$ is said to have an *oscillation of length n* if there are points $(a \leq) x_1 < \ldots < x_n (\leq b)$ and $\sigma = 1$ or -1 such that $\sigma(-1)^i (f(x_{i+1}) - f(x_i)) > 0$ $i = 1, \ldots, n - 1$.

(3) Let G be an n-dimensional subspace of $F[a, b]$. If no $u \in U$ has an oscillation of length $n + 1$, G is called an *oscillation space*.

Notations. (1) $C^1[a, b]$ and $AC[a, b]$ denote the space of all real-valued continuously differentiable functions on a compact real interval $[a, b]$ and the space of all real-valued absolutely continuous functions on $[a, b]$, respectively.

(2) For a subspace G of $AC[a, b]$, we write $G' := \{f' \mid f \in G\}$ and call G' *the space of derivatives of G.*

The main purpose of this section is to extend the following two results.

Theorem 3.2.1.(Zwick[43;(13) Corollary]) *Let G be an n-dimensional subspace of $C^1[a, b]$, which contains constants. Then the following statements are equivalent:*

(1) *G is an oscillation space.*

(2) *G has a basis $\{u_1 = 1, \ldots, u_n\}$ such that $\{u_i\}_{i=1}^n$ is a complete WT_P-system.*

(3) *G' is a WT_P-space.*

Theorem 3.2.2.(Kitahara[7;Theorem 4]) *Let G be an $n(\geq 2)$-dimensional subspace of $C^1[a, b]$, which contains constants. Then the following statements are equivalent:*

(1) *G has a basis $\{u_1 = 1, \ldots, u_n\}$ such that $\{u_i\}_{i=1}^n$ is a complete T_P-system.*

(2) *G' is an H_I-space.*

Before proceeding to the main subject, we consider Theorem 3.2.2. Let us look at the following subclass of complete T_P-systems in $C^1[a, b]$.

Definition 2. Let $\{1\} \cup \{u_i\}_{i=1}^n$ be a complete T_P-system in $C^1[a, b]$. Suppose that there exists a strictly increasing function $t = \phi(x) \in C^1[a, b]$ such that $u_i(\phi^{-1}(t))$

31

$\in C^1[\phi(a), \phi(b)]$, $i = 1, \ldots, n$, and $\{\frac{d}{dt}u_i(\phi^{-1}(t))\}_{i=1}^n$ is a complete T_P-system on $(\phi(a), \phi(b))$. Then $\{1\} \cup \{u_i\}_{i=1}^n$ is called a complete T_P-system with (DT)-property.

Proposition 3.2.3. *Let G be an $n(\geq 2)$-dimensional subspace of $C^1[a, b]$, which contains constants. Then the following statements are equivalent:*

(1) *G has a basis $\{u_1 = 1, \ldots, u_n\}$ such that $\{u_i\}_{i=1}^n$ is a complete T_P-system with (DT)-property.*

(2) *G' has a basis $\{v_1, \ldots, v_{n-1}\}$ such that*

$$v_i(x) = w(x)z_i(x), \qquad x \in [a, b] \qquad i = 1, \ldots, n-1,$$

where w is a nonnegative continuous function such that $Z(w) = \{x| \ x \in [a, b], w(x) = 0\}$ is nowhere dense in $[a, b]$ and $\{z_i\}_{i=1}^{n-1}(\subset C[a, b])$ is a complete T_P-system on (a, b).

Proof. Suppose that $\{u_1 = 1, \ldots, u_n\}$ is a complete T_P-system with (DT)-property. By Definition 2, there is a strictly increasing function $t = \phi(x) \in C^1[a, b]$ such that $u_i(\phi^{-1}(t)) \in C^1[\phi(a), \phi(b)]$, $i = 2, \ldots, n$ and $\{\frac{d}{dt}u_i(\phi^{-1}(t))\}_{i=2}^n$ is a complete T_P-system on $(\phi(a), \phi(b))$. Hence, we have

$$\frac{d}{dx}u_i(x) = \frac{d}{dx}\phi(x)\frac{d}{dt}u_i(\phi^{-1}(t)), \qquad i = 2, \ldots, n.$$

Setting $w(x) = \frac{d}{dx}\phi(x)$ and $z_i(x) = \frac{d}{dt}u_i(\phi^{-1}(t))$, (2) follows immediately.

Conversely suppose that (2) holds. Then we set a strictly increasing function $\phi \in C^1[a, b]$ such that

$$t = \phi(x) = \int_a^x w(y)dy \qquad x \in [a, b].$$

Setting $u_i(x) = w_i(\phi(x))$, $i = 2, \ldots, n$, where $w_i(t) = u_i(a) + \int_{\phi(a)}^t z_i(\phi^{-1}(t))dt$, by Theorem 3.2.2 and Definition 1, we see that $\{u_1 = 1, \ldots, u_n\}$ form a complete T_P-system with (DT)-property.

Remark 1. (1) Let E be a subspace of $F[a, b]$ and let $\{u_i\}_{i=1}^n$ be an H_I-system in E. If $\{u_i\}_{i=1}^n$ is represented as a form of (2) in Proposition 3.2.3 and if $w(x), z_i(x) \in E$, $i = 1, \ldots, n$, then we call $\{u_i\}_{i=1}^n$ *separated in E* (see Remark 1 in 3.3).

(2) There exist a lot of H_I-systems separated in $C[a, b]$ or $C^n[a, b]$. But every H_I-system $\{u_i\}_{i=1}^n$ in $C[a, b]$ or $C^n[a, b]$ is not always separated in these function spaces (see Example 2 in 3.3).

To give extensions of Theorem 3.2.1 and 3.2.2, we prepare some lemmas.

Lemma 3.2.4.(Zwick[43;(11)Theorem]) *A finite dimensional subspace G of $C[a, b]$ is an oscillation space if and only if G has a basis $\{u_1 = 1, \ldots, u_n\}$ such that $\{u_i\}_{i=1}^n$ is a complete WT_P-system.*

Lemma 3.2.5. *Let $\{u_i\}_{i=1}^n$ be a system in $C[a, b]$. $\{u_i\}_{i=1}^n$ is a WT_I-system if and only if it is a WT_P-system.*

Proof. Suppose that $\{u_i\}_{i=1}^n$ is a WT_P-system. From (2.3.4), it immediately follows that $\{u_i\}_{i=1}^n$ is a WT_I-system.

Assume that $\{u_i\}_{i=1}^n$ is not a WT_P-system, that is, there exist points $(a \leq)x_1 < \ldots < x_n(\leq b)$, $(a \leq)y_1 < \ldots < y_n(\leq b)$ such that $D\begin{pmatrix} u_1 & \cdots & u_n \\ x_1 & \cdots & x_n \end{pmatrix} > 0$ and $D\begin{pmatrix} u_1 & \cdots & u_n \\ y_1 & \cdots & y_n \end{pmatrix} < 0$. Then there are subintervals of $[a, b]$, $I_{x_1} < \ldots < I_{x_n}$, and $I_{y_1} < \ldots < I_{y_n}$ such that

$$x_i \in I_{x_i}, \quad y_i \in I_{y_i}, \quad i = 1, \ldots, n,$$

$$D\begin{pmatrix} u_1 & \cdots & u_n \\ z_1 & \cdots & z_n \end{pmatrix} > 0 \quad \text{for all } z_i \in I_{x_i}, \; i = 1, \ldots, n,$$

$$D\begin{pmatrix} u_1 & \cdots & u_n \\ w_1 & \cdots & w_n \end{pmatrix} < 0 \quad \text{for all } w_i \in I_{y_i}, \; i = 1, \ldots, n.$$

Hence, from (2.3.4), we have

$$\det\left(\int_{I_{x_j}} u_i(x)dx\right)_{i,j=1}^n > 0$$

and

$$\det\left(\int_{I_{y_j}} u_i(x)dx\right)_{i,j=1}^n < 0.$$

This means that $\{u_i\}_{i=1}^n$ is not a WT_I-system.

Let us recall that S is the set of all nondegenerate subintervals of $[a, b]$ with order $\leq$ (see 1.3).

Lemma 3.2.6. *Let $\{u_i\}_{i=1}^n$ be a system in $L^1[a, b]$ and let $v_i(I) = \int_I u_i dx$ for $I \in S$. Then the following statements are equivalent:*

(1) $\{u_i\}_{i=1}^n$ *is a WT_I-system.*

(2) *Let T be any subset of S such that T is a linearly ordered set with $\leq$ and contains at least $n+1$ elements. No $v \in [v_1, \ldots, v_n]$ has an alternation of length $n+1$ on T.*

33

Proof. $(1) \rightarrow (2)$. Let $\mathcal{T}$ be a subset of $\mathcal{S}$ satisfying the condition in (2). Since $\{u_i\}_{i=1}^n$ is a $WT_{\mathcal{I}}$-system, for any n subintervals $I_1, \ldots, I_n$ of $\mathcal{T}$,

$$D \begin{pmatrix} \sigma u_1 & \cdots & u_n \\ u_{I_1} & \cdots & u_{I_n} \end{pmatrix} \geq 0,$$

where σ is constant 1 or -1 and u_{I_j}, $j = 1, \ldots, n$, denotes a linear functional on $L^1[a, b]$ such that $u_{I_j}(f) = \int_{I_j} f(x) dx$ for all $f \in L^1[a, b]$. This means that $\{v_i\}_{i=1}^n$ is a $WT_{\mathcal{J}}$-system, where $\mathcal{J} = \{(I_1, \ldots, I_n) \mid (I_1, \ldots, I_n) \in \mathcal{T}^n, I_1 < \ldots < I_n\}$. From Proposition 1.3.2, we obtain (2).

$(2) \rightarrow (1)$. Suppose that $\{u_i\}_{i=1}^n$ is not a $WT_{\mathcal{I}}$-system. Then there exist nondegenerate subintervals of $[a, b]$ $I_1 < \ldots < I_n$ and $J_1 < \ldots < J_n$ such that

$$\det D_1 := D \begin{pmatrix} \sigma u_1 & \cdots & u_n \\ u_{I_1} & \cdots & u_{I_n} \end{pmatrix} > 0,$$

$$\det D_2 := D \begin{pmatrix} \sigma u_1 & \cdots & u_n \\ u_{J_1} & \cdots & u_{J_n} \end{pmatrix} < 0,$$

where σ is constant 1 or -1. For I_1 and J_1, $\det D_1$ and $\det D_2$ are expressed as

$$\det D_1 = \alpha_1 \int_{I_1} u_1 dx + \cdots + \alpha_n \int_{I_1} u_n dx = \int_{I_1} \sum_{k=1}^n \alpha_k u_k dx > 0,$$

$$\det D_2 = \beta_1 \int_{J_1} u_1 dx + \cdots + \beta_n \int_{J_1} u_n dx = \int_{J_1} \sum_{k=1}^n \beta_k u_k dx < 0,$$

where α_k, β_k, $k = 1, \ldots, n$, denote the cofacters of $(1, k)$ component of D_1 and D_2, respectively. Then we can take subintervals $I_1' \subset I_1$ and $J_1' \subset J_1$ such that $I_1' \cap J_1' = \emptyset$ and

$$\int_{I_1'} \sum_{k=1}^n \alpha_k u_k dx > 0, \quad \int_{J_1'} \sum_{k=1}^n \beta_k u_k dx < 0.$$

Hence, we have

$$D \begin{pmatrix} \sigma u_1 & \cdots & u_n \\ u_{I_1'} & \cdots & u_{I_n} \end{pmatrix} > 0, \quad D \begin{pmatrix} \sigma u_1 & \cdots & u_n \\ u_{J_1'} & \cdots & u_{J_n} \end{pmatrix} < 0.$$

Continuing this procedure, we obtain subintervals $I_k' \subset I_k$ and $J_k' \subset J_k$, $k = 1, \ldots, n$ satisfying that

$$I_k' \cap J_\ell' = \emptyset \qquad \text{for all} \ \ k, \ell = 1, \ldots, n,$$

$$D \begin{pmatrix} \sigma u_1 & \cdots & u_n \\ u_{I_1'} & \cdots & u_{I_n'} \end{pmatrix} > 0, \quad D \begin{pmatrix} \sigma u_1 & \cdots & u_n \\ u_{J_1'} & \cdots & u_{J_n'} \end{pmatrix} < 0.$$

If we set $\mathcal{T} = \{I_1', \ldots, I_n', J_1', \ldots, J_n'\} \subset \mathcal{S}$, then it satisfies the condition in (2). But, from the above inequalities and Proposition 1.3.2, we get a $v \in [v_1, \ldots, v_n]$ which has

34

an alternation of at least length $n+1$ on $\mathcal{T}$. Hence, (2) does not hold. This completes the proof.

Now we show

Theorem 3.2.7. *Let G be an n-dimensional subspace of $AC[a, b]$, which contains constants. Then the following statements are equivalent:*

(1) *G is an oscillation space.*

(2) *G has a basis $\{u_1 = 1, \ldots, u_n\}$ such that $\{u_i\}_{i=1}^{n}$ is a complete WT_P-system.*

(3) *G' is a $WT_{\mathcal{I}}$-space in $L^1[a, b]$.*

Proof (1) $\leftrightarrow$ (2). This immediately follows from Lemma 3.2.4.

(3) $\rightarrow$ (1). Suppose that G' is a $WT_{\mathcal{I}}$-space in $L^1[a, b]$ and G is not an oscillation space. There are $n + 1$ points $(a \leq)x_1 < \ldots < x_{n+1}(\leq b)$ and a $u \in G$ such that $(-1)^i(u(x_{i+1}) - u(x_i)) > 0$, $i = 1, \ldots, n$, i.e., $(-1)^i \int_{x_i}^{x_{i+1}} u'(x)dx > 0$, $i = 1, \ldots, n$. Since the dimension of G' is $n - 1$, from Lemma 3.2.6, we have a contradiction to (1).

(1) $\rightarrow$ (3). Assume that G is an oscillation space and that G' is not a $WT_{\mathcal{I}}$-space in $L^1[a, b]$. By Lemma 3.2.6, there exist a $u' \in G'$ and n subintervals $[a_1, b_1] < \cdots < [a_n, b_n]$ of $[a, b]$ satisfying $(-1)^k \int_{[a_k, b_k]} u'(x)dx > 0$, $k = 1, \ldots, n$, i.e., $(-1)^k(u(b_k) - u(a_k)) > 0$, $k = 1, \ldots, n$. Then we take $n + 1$ points $\{x_i\}_{i=1}^{n+1}$ of $[a.b]$ such that

$$x_1 = a_1 \ , \qquad\qquad x_{n+1} = b_n,$$

$$x_i \in \{x \mid x \in [a_{i-1}, a_i], \ (-1)^i u(x) = \max_{y \in [a_{i-1}, a_i]}(-1)^i u(y) \ \}, \qquad i = 2, \ldots, n.$$

$\{x_i\}_{i=1}^{n+1}$ satisfies $x_1 < \cdots < x_{n+1}$ and $(-1)^k (u(x_{k+1}) - u(x_k)) > 0$, $k = 1, \ldots, n$. This contradicts the fact that G is an oscillation space. Hence, we prove our conclusion.

Next we show

Theorem 3.2.8. *Let G be an $n(\geq 2)$-dimensional subspace of $AC[a, b]$, which contains constants. Then the following statements are equivalent:*

(1) *G has a basis $\{u_1 = 1, \ldots, u_n\}$ such that $\{u_i\}_{i=1}^{n}$ is a complete T_P-system.*

(2) *G' is an $H_{\mathcal{I}}$-space in $L^1[a, b]$.*

Proof. (1) $\rightarrow$ (2). Suppose that G is a complete T_P-space with a basis $\{u_1 = 1, \ldots, u_n\}$ and suppose that G' is not an $H_{\mathcal{I}}$-space in $L^1[a, b]$. Then there exist subintervals $I_1 < \cdots < I_{n-1}$ of $[a, b]$ such that

$$D \begin{pmatrix} u'_2 & \cdots & u'_n \\ u_{I_1} & \cdots & u_{I_{n-1}} \end{pmatrix} = 0,$$

35

where each u_{I_k}, $k = 1, \ldots, n-1$, denotes a linear functional on $L^1[a, b]$ which appeared in the proof in Lemma 3.2.6. From this, we have a $u' \in [u_2', \ldots u_n'] - \{0\}$ satisfying

$$(3.2.1) \qquad \int_{I_k} u' dx = 0, \qquad k = 1, \ldots, n-1.$$

Since $\{u_i\}_{i=1}^n$ is a T_P-system, every function of $[u_2', \ldots, u_{n-1}'] - \{0\}$ does not vanish identically on any nondegenerate subinterval on $[a, b]$. From (3.2.1), we have subintervals P_k, N_k of I_k, $k = 1, \ldots, n-1$, such that

$$(3.2.2) \qquad P_k \cap N_k = \emptyset$$
$$\int_{P_k} u' dx > 0, \qquad \int_{N_k} u' dx < 0.$$

By (3.2.2), there exist n subintervals $J_1 < \ldots < J_n$ of $[a, b]$ satisfying $(-1)^k \int_{J_k} u' dx > 0$, $k = 1, \ldots, n$. Hence, by the proof of (1) $\rightarrow$ (3) in Theorem 3.2.7, G is not an oscillation space, but this contradicts our hypothesis of G.

(2) $\rightarrow$ (1). Suppose that G' is an $H_{\mathcal{I}}$-space. Since every $H_{\mathcal{I}}$-space is a $WT_{\mathcal{I}}$-space, from Theorem 3.2.7, we have a basis $\{u_1 = 1, \ldots, u_n\}$ of G such that $\{u_i\}_{i=1}^n$ is a complete WT_P-system. Furthermore we shall show that each system $\{u_i\}_{i=1}^j$, $2 \leq j \leq n$, is a T_P-system.

Suppose that a function $f = \Sigma_{i=1}^j a_i u_i \in [u_1, \ldots, u_j] - \{0\}$ vanishes identically on a subinterval $[c, d]$ of $[a, b]$. Since $a_2^2 + \cdots + a_j^2 \neq 0$, f' is contained in $G' - \{0\}$ and vanishes identically on $[c, d]$. But this is contradictory to our assumption on G'. Hence, applying Lemma 2.4.12, we obtain the fact that each $\{u_i\}_{i=2}^j$, $2 \leq j \leq n$, is a T_P-system on the open interval (a, b). By Theorem 3.2.7, $\{u_i'\}_{i=2}^j$ is a $WT_{\mathcal{I}}$-system on $[a, b]$. If $u \in [u_1, \ldots, u_j] - \{0\}$ has j zeros $(a \leq) z_1 < \ldots < z_j (\leq b)$ in $[a, b]$, then, there are subintervals P_k, N_k of (z_k, z_{k+1}), $k = 1, \ldots, j-1$, satisfying the conditions (3.2.2). There exist j subintervals $J_1 < \ldots < J_j$ of $[a, b]$ such that $(-1)^k \int_{J_k} u' dx > 0$, $k = 1, \ldots, j$. By Lemma 3.2.6, this contradicts the fact that $\{u_i'\}_{i=2}^j$ is a $WT_{\mathcal{I}}$-system on $[a, b]$. Hence, every $u \in [u_1, \ldots, u_j] - \{0\}$ has at most $j - 1$ zeros in $[a, b]$. This completes the proof.

Remark 2. The readers can easily see that, by Lemma 3.2.5, Theorem 3.2.7 and 3.2.8 are extensions of Theorem 3.2.1 and 3.2.2, respectively.

3.3 Representations of WT-Systems in $F[a,b]$

It is well known that integral representations of complete T_P or complete WT_P-systems play a main part in the representation of WT-systems. First we present some of their results.

Theorem 3.3.1.(Karlin and Studden[6;Theorem 1.2 in Chap. XI]) *Let $u_0,\ldots,u_n$ be a system in $C^n[a,b]$ satisfying*

$$u_k^{(p)}(a) = 0, \qquad p = 0,\ldots,k-1, \qquad k = 1,\ldots,n.$$

The following three statements are equivalent:

(1) $\{u_i\}_{i=0}^n$ *is a complete $T_{\widetilde{P}}$-system.*

(2) $W(u_0,\ldots,u_k)(t) > 0$ *for $k = 0,\ldots,n$ and all $t \in [a,b]$, where $W(u_0,\ldots,u_k)(t)$ denotes the Wronskian determinant of $\{u_i\}_{i=0}^k$ at $t \in [a,b]$.*

(3) $\{u_i\}_{i=0}^n$ *has a representation of the form (3.3.1),*

$$u_0(t) = w_0(t)$$

$$u_1(t) = w_0(t)\int_a^t w_1(x_1)dx_1$$

$$u_2(t) = w_0(t)\int_a^t w_1(x_1)\int_a^{x_1} w_2(x_2)dx_2dx_1$$

(3.3.1)
$$\vdots$$

$$u_n(t) = w_0(t)\int_a^t w_1(x_1)\int_a^{x_1} w_2(x_2)\cdots\int_a^{x_{n-1}} w_n(x_n)dx_n\cdots dx_1,$$

where each w_i, $i = 0,\ldots,n$, is a strictly positive function in $C^{n-i}[a,b]$.

Condition (2) in Theorem 3.3.1 is a necessary and sufficient condition so that systems $\{u_i\}_{i=0}^n$ in $C^n[a,b]$ are complete $T_{\widetilde{P}}$-systems. But if we consider a weaker condition such that $\{t \mid W(u_0,\ldots,u_k)(t) > 0, t \in [a,b]\}, k = 0,\ldots,n$, is dense in $[a,b]$, then it does not always follow from this condition that $\{u_i\}_{i=0}^n$ is a WT_P-system.

Example 1. Let us consider the following system $u_0, u_1 \in C^1[-2,2]$:

$$u_0(x) = x^2,$$

$$u_1(x) = \begin{cases} x^2(x+1) & x \in [-2,0] \\ x^2(x-1) & x \in (0,2], \end{cases}$$

Since $W(u_0)(t) = t^2$ and $W(u_0,u_1)(t) = t^4$, $W(u_0)(t)$ and $W(u_0,u_1)(t)$ are positive except at 0. But u_1 changes sign three times on $[-2,2]$. Hence, $\{u_0,u_1\}$ is not a WT_P-system.

37

When systems $\{u_i\}_{i=0}^n$ in $C^n[a, b]$ have representations like (3.3.1), we obtain

Proposition 3.3.2. *Let $u_0, \ldots, u_n$ be a system in $C^n[a, b]$. Suppose that $\{u_i\}_{i=0}^n$ has a representation of (3.3.1) and suppose that each $w_i(x) \in C^{n-i}[a, b]$, $i = 0, \ldots, n$ is strictly positive on $[a, b]$ except a nowhere dense subset of $[a, b]$. Then $\{u_i\}_{i=0}^n$ is a complete $H_\mathcal{I}$-system such that each $\{t \mid W(u_0, \ldots, u_k)(t) > 0, \ t \in [a, b]\}, k = 0, \ldots, n$, is dense in $[a, b]$. In particular, $\{u_i/u_0\}_{i=1}^n$ is a complete T_P-system in $C^n[a, b]$ (each u_i/u_0 is an extended function on $[a, b]$).*

Proof. We prove this by induction on n. We easily observe that this is true for $n = 0$ and $W(u_0)(t) = u_0(t) = w_0(t) \ t \in [a, b]$. Suppose that the results hold for $n = k - 1 \ (k \geq 1)$ and that $W(u_0, \ldots, u_{k-1}) = w_0^k w_1^{k-1} \cdots w_{k-1}$. Let us consider the case $n = k$.

For $\{u_i\}_{i=0}^k$ in $C^k[a, b]$ with the condition of this proposition, $\{u_0/u_0 = 1, u_1/u_0, \ldots, u_k/u_0\}$ (each u_i/u_0 is an extended function on $[a, b]$) is a system in $C^k[a, b]$ such that $\{(u_1/u_0)', \ldots, (u_k/u_0)'\}$ has a representation of the form (3.3.1). Appling the Leibniz rule of differentiation to $\dfrac{u_i}{u_0} \in C^k[a, b]$, $i = 1, \ldots, k$, we obtain

$$(\frac{u_i}{u_0})^{(r)}(t) = \Sigma_{j=0}^r \binom{r}{j} u_i^{(r-j)}(t) (\frac{1}{u_0})^{(j)}(t), \qquad r = 1, \ldots, k, \qquad t \in [a, b] - Z(u_0).$$

Using this, we verify that (see p.377 in Karlin and Studden[6])

$$W(u_0, \ldots, u_k)(t) = u_0^{k+1} W((\frac{u_1}{u_0})', \ldots, (\frac{u_k}{u_0})') \qquad t \in [a, b].$$

By the hypothesis of induction, $\{(u_1/u_0)', \ldots, (u_k/u_0)'\}$ is a complete $H_\mathcal{I}$-system in $C^{k-1}[a, b]$ and

(3.3.2) $$W(u_0, \ldots, u_k) = w_0^{k+1} w_1^k \cdots w_k.$$

From Theorem 3.2.2 and Proposition 2.3.1, it follows that $\{u_0/u_0 = 1, u_1/u_0, \ldots, u_k/u_0\}$ is a complete T_P-system and $\{u_i\}_{i=0}^n$ is an $H_\mathcal{I}$-system. By (3.3.2), it is clear that $\{t \mid W(u_0, \ldots, u_k)(t) > 0, \ t \in [a, b]\}$ is dense in $[a, b]$. Hence, the results hold for $n = k$.

Let M be a subset of the real line and let $\{u_i\}_{i=1}^n$ be a WT_P-system in $F(M)$ and let $G = [u_1, \ldots, u_n]$. If the spaces $G|_{(-\infty, c) \cap M}$ and $G|_{(c, \infty) \cap M}$ have the same dimension as G for all $c \in M$, then $\{u_i\}_{i=1}^n$ is called *nondegenerate*.

Here, we introduce a result of integral representations of nondegenerate complete WT_P-system. The readers can obtain more detailed results or further investigations of this topic in Rutman[23], Schwenker[25], Zalik[35,36,37,38] and Zielke[42].

38

Theorem 3.3.3.(Zielke[42;Theorem 3]) *Let M be a subset of the real line and let $\{u_0 = 1, u_1, \ldots, u_n\}$ be a nondegenerate complete WT_P-system. Then there exist a basis $\{v_0, v_1, \ldots, v_n\}$ of $[u_0, \ldots, u_n]$, a strictly increasing function h on M, continuous increasing functions w_i, $i = 1, \ldots, n$, on $I = (\inf h(M), \sup h(M))$ and $c \in I$ such that, for all $x \in M$,*

$$v_0(x) \equiv 1$$

$$v_1(x) = \int_c^{h(x)} dw_1(t_1)$$

$$v_2(x) = \int_c^{h(x)} \int_c^{t_1} dw_2(t_2) dw_1(t_1)$$

$$\vdots$$

$$v_n(x) = \int_c^{h(x)} \int_c^{t_1} \cdots \int_c^{t_{n-1}} dw_n(t_n) \cdots dw_1(t_1).$$

The purpose of this section is to give separated representations of WT_P-systems in general and WT_P-systems of m times continuously differentiable functions.

We treat systems $\{u_i\}_{i=1}^n$ of real-valued functions such that they consist of linearly independent functions and so that $V(\{u_i\}) = \{x | u_i(x) = 0, x \in [a, b], i = 1, \ldots, n\}$ is nowhere dense in $[a, b]$. We call $\{u_i\}$ a *nonvanishing system* if for any $x \in [a, b]$ there exists an i_0 such that $u_{i_0}(x) \neq 0$.

Before stating a theorem, we show the following lemma.

Lemma 3.3.4. *Let $\{u_i\}_{i=1}^n$ be a WT_P-system in $F[a, b]$ and let $u(x) = \max_i |u_i(x)|, x \in [a, b]$. For each $x_0 \in V(\{u_i\}) = V$, $\lim_{\substack{x \to x_0 - \\ x \in [a,b]-V}} u_i(x)/u(x)$ and $\lim_{\substack{x \to x_0 + \\ x \in [a,b]-V}}$ $u_i(x)/u(x)$, $i = 1, \ldots, n$, exist, where if $x_0 = a$ or b, the possible case is considered.*

Proof. First we shall show that, for each $x_0 \in V$, there exist a $\delta > 0$ and $i, j \in N$ with $1 \leq i, j \leq n$ such that

(3.3.3) $u(x) = |u_i(x)| \quad x \in (x_0 - \delta, x_0)$

(3.3.4) $u(x) = |u_j(x)| \quad x \in (x_0, x_0 + \delta),$

where if $x_0 = a$ or b, the possible case is considered.

Suppose (3.3.3) does not hold. Then there are monotone increasing sequences $\{x_n\}, \{y_n\}$ which converge to x_0 and satisfy

(3.3.5) $u(x_n) = |u_i(x_n)|, \quad u(y_n) = |u_j(y_n)| \quad i \neq j$

Hence either $u_i + u_j$, or $u_i - u_j$ changes sign infinitely many times on (a, x_0). This is a contradiction. In an analogous way, we can easily show the validity of (3.3.4).

39

Now we will prove the existence of $\lim\limits_{\substack{x\to x_0- \\ x\in[a,b]-V}} u_k(x)/u(x), k = 1,\ldots,n$. As for the

existence of $\lim\limits_{\substack{x\to x_0+ \\ x\in[a,b]-V}} u_k(x)/u(x)$, an analogous way can be applied. Suppose that

$\lim\limits_{\substack{x\to x_0- \\ x\in[a,b]-V}} u_k(x)/u(x)$ does not exist. Since $0 \le |u_k(x)/u(x)| \le 1$ if $x \in [a,b] - V$, there

are monotone increasing sequencecs $\{z_n\}, \{w_n\} \subset [a,b] - V$ such that

$$(3.3.6) \qquad \lim_n z_n = \lim_n w_n = x_0$$

$$(3.3.7) \qquad \lim_n u_k(z_n)/u(z_n) = \alpha < \lim_n u_k(w_n)/u(w_n) = \beta.$$

By (3.3.3), without loss of generality, we can set $u(x) = |u_i(x)|$. Hence $0.5(\alpha+\beta)u_i-u_k$ changes sign infinitely many times on (a, x_0). This contradicts the fact that $\{u_i\}$ is a WT_P-system. This completes the proof.

Now we obtain

Theorem 3.3.5. *Let* $\{u_i\}_{i=1}^n$ *be a* WT_P-*system in* $F[a,b]$ *and let* $u(x) = \max\limits_i |u_i(x)|$ $x \in [a,b]$. *Then each* $u_i, i = 1,\ldots,n$, *is represented as*

$$u_i(x) = u(x)v_i(x) \qquad x \in [a,b],$$

where $\{v_i\}_{i=1}^n$ *is a nonvanishing* WT_P-*system in* $F[a,b]$. *When* $\{u_i\}_{i=1}^n$ *is a* WT_P-*system such that any nonzero linear combination of* $\{u_i\}_{i=1}^n$ *does not vanish on a nondegenerate subinterval, the system* $\{v_i\}_{i=1}^n$ *is a* T_P-*system on* (a,b).

Proof. By Lemma 3.3.4, we can set, for each $i, i = 1,\ldots,n$,

$$v_i(x_0) = \begin{cases} u_i(x_0)/u(x_0) & x_0 \in [a,b] - V, \\ \lim\limits_{\substack{x\to x_0- \\ x\in[a,b]-V}} u_i(x)/u(x) & x_0 \in (a,b] \cap V, \\ \lim\limits_{\substack{x\to x_0+ \\ x\in[a,b]-V}} u_i(x)/u(x) & if \ x_0 = a \in V. \end{cases}$$

We easily see that the system $\{v_i\}$ is nonvanishing. Suppose that some linear combination v of $\{v_i\}$ changes sign n times on $[a,b]$, that is, for $(n+1)$ points $\{x_i\}_{i=1}^{n+1}$ in $[a,b]$, $a \le x_1 < \cdots < x_{n+1} \le b$, and for $\sigma = 1$ or -1,

$$(3.3.8) \qquad \sigma(-1)^i v(x_i) > 0 \qquad i = 1,\ldots,n+1.$$

By (3.3.8) and the definition of $\{v_i\}$, for each $x_i, i = 1,\ldots,n+1$, we can find a y_i which belongs to $([a,b] - V) \cap [x_i - \epsilon, x_i + \epsilon]$, $\epsilon = \min\limits_{1\le i\le n} |x_i - x_{i+1}|/3$, and satisfies $v(x_i)/|v(x_i)| = v(y_i)/|v(y_i)|$. Hence, from (3.3.8), we have

$$\sigma(-1)^i v(y_i)u(y_i) > 0 \qquad i = 1,\ldots,n+1.$$

40

But this contradicts the hypothesis of $\{u_i\}$. Thus, $\{v_i\}$ is a nonvanishing WT_P-system.

Let $\{u_i\}_{i=1}^n$ be a WT_P-system such that any nonzero linear combination of $\{u_i\}_{i=1}^n$ does not vanish on a nondegenerate subinterval. Suppose that a nonzero linear combination w of $\{v_i\}_{i=1}^n$, which is defined above, has n zeros in (a, b). Since $\{v_i\}$ is a nonvanishing WT_P-system, by Lemma 2.4.12, w vanishes on a nondegenerate subinterval. But the factorization of $u_i(x)$ leads to a contradiction of the hypothesis on the WT_P-system $\{u_i\}$. Hence, $\{v_i\}$ is a T_P-system on (a, b).

Example 2. Let us consider the following WT_P-system $\{u_1, u_2\}$ on $[-1, 1]$:

$$u_1(x) = \begin{cases} x^2/(x+1) & x \in [0, 1] \\ x^2 & x \in [-1, 0) \end{cases}$$

$$u_2(x) = \begin{cases} x^2 & x \in [0, 1] \\ x^3 & x \in [-1, 0). \end{cases}$$

Since $u(x) = \max_{i=1,2} |u_i(x)| = x^2$, we obtain a T_P-system $\{v_1, v_2\}$ defined in Theorem 3.3.5 such that

$$v_1(x) = \begin{cases} 1/(x+1) & x \in [0, 1] \\ 1 & x \in [-1, 0) \end{cases}$$

$$v_2(x) = \begin{cases} 1 & x \in (0, 1] \\ x & x \in [-1, 0] \end{cases}$$

By the above example, the T_P-system $\{v_i\}$ which is defined in Theorem 3.3.5 for a WT_P-system $\{u_i\}$ of continuous functions does not always consist of continuous functions.

If a WT_P-system satisfies the following condition, then we have

Theorem 3.3.6. *Let $\{u_i\}_{i=1}^n$ be nondegenerate on (a, b) and a WT_P-system in $C^m[a, b]$ such that $\dim[u_1, \ldots, u_n]|_{(a,b)} = n$ and $V = V(\{u_i\}_{i=1}^n)$ is finite. Suppose that, for each $c \in V$, there exist i_c ($1 \le i_c \le n$) and k_c ($1 \le k_c \le m$) such that $u_{i_c}^{(k_c)}(c) \neq 0$. Then $\{u_i\}_{i=1}^n$ is a nonvanishing WT_P-system on (a, b) or is represented as*

$$u_i(x) = (x - a_1)^{2p_1} \cdots (x - a_k)^{2p_k} v_i(x) \quad i = 1, \ldots, n, \quad x \in (a, b),$$

where $\{a_j\}_{j=1}^k = V \cap (a, b)$, $p_i \in N$ and $\{v_i\}_{i=1}^n$ is a system in $C[a, b]$ and a nonvanishing WT_P-system on (a, b). Moreover, if $\{u_i\}_{i=1}^n$ is an H_I-system, $\{v_i\}_{i=1}^n$ is a system in $C[a, b]$ and a T_P-system on (a, b).

41

Proof. If $V \cap (a, b)$ is empty, $\{u_i\}_{i=1}^{n}$ itself is a nonvanishing WT_P-system on (a, b).

Suppose that $V \cap (a, b) = \{a_i\}_{i=1}^{k}, a < a_1 < \cdots < a_k < b$. For each a_i and each j, $j = 1, \ldots, n$, we set $m(i, j) = r$ if u_j is expressed as $u_j(x) = \alpha(x - a_i)^r + o((x - a_i)^r)$, $\alpha \neq 0$, $1 \leq r \leq m$, and set $m(i, j) = \infty$ if $u_j'(a_i) = \cdots = u_j^{(m)}(a_i) = 0$. By the hypothesis, $m(i, j)$, $i = 1, \ldots, k$, $j = 1, \ldots, n$ are well defined.

When we put $q_i = \min\{m(i, 1), \ldots, m(i, n)\}$ for $i = 1, \ldots, k$, it is shown that every q_i is even. Suppose to the contrary that, say, q_1 is odd. Set $w_i(x) = u_i(x)/(x - a_1)^{q_1}, i = 1, \ldots, n$. Clearly each w_i is continuous on $[a, b]$. Since $\{u_i\}_{i=1}^{n}$ is a nondegenerate WT_P-system on (a, b), the dimension of $[w_1, \ldots, w_n]|_{(a, a_1)}$ is n. Hence, there exists a nonzero linear combination w_0 of $\{w_i\}$ such that, for some n points $\{b_i\}_{i=1}^{n}$ in $(a, a_1), a < b_1 < \cdots < b_n < a_1$,

(3.3.9) $\qquad (-1)^i w_0(b_i) > 0 \qquad i = 1, \ldots, n$

(3.3.10) $\qquad w_0(a_1) \neq 0$.

By (3.3.9), (3.3.10) and the assumption that q_1 is odd, the linear combination $v_0(x) = (x - a_1)^{q_1} w_0(x)$ of $\{u_i\}$ changes sign at least n times on (a, b), but this contradicts the fact that $\{u_i\}$ is a WT_P-system. As for the other $q_i's$, it is proven analogously. Hence, if we set $2p_i = q_i, i = 1, \ldots, k$ and $v_i(x) = u_i(x)/[(x - a_1)^{2p_1} \cdots (x - a_k)^{2p_k}], i = 1, \ldots, n$, then $\{v_i\}$ is a system in $C[a, b]$ and nonvanishing WT_P-system on (a, b).

Corollary 3.3.7. *Let $\{u_i\}_{i=1}^{n}$ be a WT_P-system of analytic functions on $[a, b]$, i.e., $\{u_i\}_{i=1}^{n}$ is a WT_P-system and each $u_i, 1 \leq i \leq n$, is analytic on an open interval containing $[a, b]$. Then $\{u_i\}$ is a T_P-system on (a, b) or is represented as*

$$u_i(x) = (x - a_1)^{2p_1} \cdots (x - a_k)^{2p_k} v_i(x) \qquad i = 1, \ldots, n, \quad x \in (a, b),$$

where $\{a_j\}_{j=1}^{k} = V \cap (a, b)$, $p_i \in N$ and $\{v_i\}$ is a T_P-system of analytic functions on (a, b).

Proof. Since each $u_i, i = 1, \ldots, n$, is analytic on $[a, b]$, we easily observe that $\{u_i\}_{i=1}^{n}$ satisfies the conditions in Theorem 3.3.6. In particular, $\{u_i\}_{i=1}^{n}$ is an H_I-system, because each $u \in [u_1, \ldots, u_n] - \{0\}$ has at most finite zeros. Hence, by Theorem 3.3.5 and the proof of Theorem 3.3.6, $\{v_i\}_{i=1}^{n}$ is a T_P-system of analytic functions on (a, b).

Remark 1. Complete H_I-systems $\{u_i\}_{i=1}^{n}$ which appeared in Theorem 3.3.6, Proposition 3.3.2, and Corollary 3.3.7 are separated in $C[a, b]$, $C^n[a, b]$ and $C^\omega[a, b]$

42

(=the space of real-valued analytic functions on $[a, b]$), respectively. Hence, the system $\{1, \int_a^x u_1 dx, \ldots, \int_a^x u_n dx\}$ is a complete T_P-system with (DT)-property.

In the rest of this section, using the methods in Gierz and Shekhtman[2], we give slight consideration to the density of specifc infinite $H_{\mathcal{I}}$-spaces.

Definition. (1) Let w be a nonnegative function in $C[a, b]$ such that $\mu(Z(w)) = 0$, where μ denotes the Lebesgue measure on $[a, b]$. Let $\{u_i\}_{i=1}^\infty$ be an infinite complete T_P-system in $C[a, b]$. For the convenience, we call each $\{wu_i\}_{i=1}^k$, $k \in N$, an $\widetilde{H}_{\mathcal{I}}$-system. If a subspace E of $C[a, b]$ contains $\{wu_i\}_{i=1}^\infty$, then E is said to be an $\widetilde{H}$-space.

(2) Let $L^p[a.b]$, $1 \le p < \infty$, be the space of all real-valued functions f on $[a, b]$ such that $|f|^p$ is integrable. Let $L^\infty[a, b]$ be the space of all real-valued measurable functions g on $[a, b]$ such that g is μ-essentially bounded. Each $L^p[a, b]$, $1 \le p \le \infty$, is endowed with the L^p-norm:

$$\|f\|_p = \begin{cases} (\int_a^b |f|^p dx)^{1/p} & (1 \le p < \infty) \\ \\ \operatorname*{ess.\,sup}_{x \in [a,b]} |f(x)| & (p = \infty) \end{cases}$$

(3) $f \in L^1[a, b]$ is said to change its sign $k(\ge 1)$ times if there exist $k - 1$ points, $(a = x_0 <)x_1 < \ldots < x_{k-1}(< x_k = b)$, and $\sigma = 1$ or -1 such that $(-1)^i \sigma f(x) \ge 0$ a.e. $x \in [x_i, x_{i+1}]$ and $(-1)^i \int_{x_i}^{x_{i+1}} \sigma f(x) dx > 0$, $i = 0, \ldots, k - 1$.

Remark 2. Systems which appeared in Proposition 3.3.2 , Theorem 3.3.6 and Corollary 3.3.7 are examples of $\widetilde{H}_{\mathcal{I}}$-system.

If $C[a, b]$ is endowed with the L^p-norm $(1 \le p < \infty)$, the dual $(C[a, b])'$ is identical with $L^q[a, b]$, where $1/p + 1/q = 1$ if $1 < p < \infty$, and $q = \infty$ otherwise. Each $f \in L^q[a, b]$ corresponds to the linear functional $\hat{f}$ on $C[a, b]$ such that $\hat{f}(g) = \int_a^b fg dx$, $g \in C[a, b]$.

Lemma 3.3.8. (cf. Gierz and Shekhtman[2;Proposition 6]) *Let $\{u_i\}_{i=1}^n$ be an $\widetilde{H}_{\mathcal{I}}$-system in $C[a, b]$. Let $f \in L^1[a, b] - \{0\}$ be a function such that $\hat{f}(g) = \int_a^b fg dx = 0$ for all $g \in [u_1, \ldots, u_n]$. If f changes sign finitely many times, then f changes sign at least $n + 1$ times.*

Proof. Suppose that f changes sign $k(\le n)$ times. Let $(a = x_0) < x_1 < \ldots < x_{k-1}(< x_k = b)$ and σ be a correspoinding set and its number, respectively. Since $\{u_i\}_{i=1}^n$ be an $\widetilde{H}_{\mathcal{I}}$-system, there exists a $u \in [u_1, \ldots, u_n]$ such that $(-1)^i u(x) > 0$ a.e.

43

on (x_i, x_{i+1}), $i = 0, \ldots, k-1$. Hence, we have

$$\hat{f}(u) = \int_a^b f u \, dx = \sum_{i=0}^{k-1} \int_{x_i}^{x_{i+1}} f u \, dx > 0.$$

But this contradicts the hypothesis of f.

Lemma 3.3.9. (cf. Gierz and Shekhtman[2;Theorem 8]) *Let G be an $\widetilde{H}$-space of $C[a,b]$ and let f be a function in $L^1[a,b]$ such that $\int_a^b f g \, dx = 0$ for all $g \in G$. Then f does not change sign finitely many times.*

Proof. Suppose that f changes sign finitely many times, for example k times. Since G is an $\widetilde{H}$-space, G contains a k dimensional $\widetilde{H}_I$-space. By Lemma 3.3.8, f must change sign at least $k+1$ times, which leads to a contradiction.

Combining the proof of Theorem 10 in Gierz and Shekhtman[2] with the above two lemmas, we can show the following result without difficulty.

Theorem 3.3.10. (cf. Gierz and Shekhtman[2;Thereom 10]) *Let G be an $\widetilde{H}$-space of $C[a,b]$. Let*

$$T : (C[a,b], \| \cdot \|_p) \to (C[a,b], \| \cdot \|_p), \quad 1 \le p < \infty,$$

be a linear operator such that

$$T(f)(x) = \int_a^b k(s,x) f(s) \, ds, \qquad f \in C[a,b],$$

where $k(s,x)$ satisfies

(i) $k(s,x)$ is analytic on $[a,b] \times [a,b]$,

(ii) $\mathrm{Span}\{k(s,x) | s \in [a,b]\}$ is dense in $(C[a,b], \| \cdot \|_p)$.

Then $T(G)$ is dense in $(C[a,b], \| \cdot \|_p)$.

When $C^1[a,b]$ is endowed with the following norm:

$$\|f\|_v = |f(a)| + \int_a^b |f'| \, dx \qquad f \in C^1[a,b],$$

then we have

Corollary 3.3.11. *Let G be an infinite complete T_P-space of $C^1[a,b]$ such that G' is an $\widetilde{H}$-space. Let*

$$T : (C^1[a,b], \| \cdot \|_v) \to (C^1[a,b], \| \cdot \|_v)$$

be a linear operator such that

$$T(f)(x) = \int_a^x \int_a^b k(s,v) f'(s) \, ds \, dv + f(a),$$

where $k(s,x)$ satisfies

(i) $k(s,v)$ *is analytic on* $[a,b] \times [a,b]$,

(ii) $\text{Span}\{k(s,v)|s \in [a,b]\}$ *is dense in* $C[a,b]$ *with the* L^1*-norm.*

Then $T(G)$ is dense in $(C^1[a,b], \|\cdot\|_v)$.

3.4 Adjoined Functions to $T_{\mathcal{P}}$ or $H_{\mathcal{I}}$-Spaces

First we give a definition of adjoined functions to a subspace of $F[a,b]$ or $F(a,b)$ ($F(a,b)$ denotes the space of all real-valued functions on an open real interval).

Definition. Let $\{u_i\}_{i=1}^n$ be an X-system in $F[a,b]$ or $F(a,b)$, where an X-system is one of the systems given in Definition 1 in 1.2. If there exists a function f in $F[a,b]$ or $F(a,b)$ such that $[u_1, \ldots, u_n, f]$ is an X-space, then f is called an *adjoined function to the X-space* $[u_1, \ldots, u_n]$.

The following results by Zielke[40,41] and Zalik[33] are fundamental to the study of adjoined functions.

Theorem 3.4.1. (1) *Let $\{u_i\}_{i=1}^n$ be a complete $T_{\mathcal{P}}$-system in $F(a,b)$. Then there exists a $u_{n+1} \in F(a,b)$ which is adjoined to the complete $T_{\mathcal{P}}$-space $[u_1, \ldots, u_n]$.*

(2) *Let $\{u_i\}_{i=1}^n$ be a $T_{\mathcal{P}}$-system in $C[a,b]$. Then there exists a u_{n+1} in $C[a,b]$ which is adjoined to the $T_{\mathcal{P}}$-space $[u_1, \ldots, u_n]$.*

We recall that, for a system $\{u_i\}_{i=1}^n$, the convexity cone $K[u_1, \ldots, u_n]$ is the set of all real-valued functions f defined on (a,b) for which the determinants

$$D\begin{pmatrix} u_1 & \cdots & u_n & f \\ t_1 & \cdots & t_n & t_{n+1} \end{pmatrix} \geq 0,$$

whenever $a < t_1 < \cdots < t_n < t_{n+1} < b$. Furthermore, we denote by $K_c[u_1, \ldots, u_n]$ the set of all functions in $K[u_1, \ldots, u_n]$ which are continuous on $[a,b]$. In this section, we use the notation $\mathcal{P}_n$, $n \in N$ for the subset $\{(x_1, \ldots, x_n) \mid (a \leq)x_1 < \cdots < x_n(\leq b)\}$ of $[a,b]^n$.

The proof of Theorem 1 in Zalik[33] shows the existence of a function in K_c which satisfies the following conditions.

Lemma 3.4.2. *Let $\{u_i\}_{i=1}^n$ be a $T_{\mathcal{P}}$-system (resp. $H_{\mathcal{I}}$-system) in $C[a,b]$. For any n points $(a \leq)x_1 < \cdots < x_n(\leq b)$, there exists an $f \in K_c[u_1, \ldots, u_n]$ such that*

45

(3.4.1) $f(x) = \begin{cases} 0, & x \in [a, x_n] \\ \sum_{i=1}^n a_i u_i(x), & x \in [x_n, b], \text{ where } a_n = 1, \end{cases}$

and

(3.4.2) $f(x) > 0$ for all $x \in (x_n, b]$

(resp.   (3.4.1)   and

(3.4.3) $\{x \mid f(x) > 0, x \in (x_n, b]\}$ is dense in $(x_n, b]$).

By Lemma 1 in Zalik[33] and Lemma 3.4.2, we can easily show the existence of adjoined functions to T_P-spaces of $AC[a, b]$ or H_I-spaces of $C[a, b]$.

Theorem 3.4.3. (1) *Every finite dimensional T_P-space of $AC[a, b]$ has an adjoined function which is absolutely continuous on $[a, b]$.*

(2) *Every finite dimensional H_I-space of $C[a, b]$ has an adjoined function which is continuous on $[a, b]$.*

Proof. (1) Let $\{u_i\}_{i=1}^n$ be a T_P-system in $AC[a, b]$ and let $\{x_i\}_{i=1}^{n+1}$ be any $n + 1$ points in $[a, b]$ with $x_1 < \cdots < x_{n+1}$. For the n points $\{x_i\}_{i=1}^n$, using Lemma 3.4.2, we have an $f \in K_c[u_1, \ldots, u_n]$ satisfying (3.4.1) and (3.4.2). Hence, we get

$$D \begin{pmatrix} u_1 & \cdots & u_n & f \\ x_1 & \cdots & x_n & x_{n+1} \end{pmatrix} = f(x_{n+1}) \cdot D \begin{pmatrix} u_1 & \cdots & u_n \\ x_1 & \cdots & x_n \end{pmatrix} > 0.$$

Since $u_i, i = 1, \ldots, n$, and f are continuous, there is a neighbourhood $U_{\mathbf{x}}$ of $\mathbf{x} = (x_1, \ldots, x_{n+1})$ such that

$$D \begin{pmatrix} u_1 & \cdots & u_n & f \\ y_1 & \cdots & y_n & y_{n+1} \end{pmatrix} > 0 \quad \text{for all } (y_1, \ldots, y_{n+1}) \in U_{\mathbf{x}}.$$

Since the family $\{U_{\mathbf{x}}\}_{\mathbf{x} \in \mathcal{P}_{n+1}}$ is a covering of $\mathcal{P}_{n+1}$, there is a subcovering $\{U_{\mathbf{x_i}}\}_{i=1}^\infty$. Let $f_i, i \in N$, be a corresponding function in $K_c[u_1, \ldots, u_n]$ to $\mathbf{x_i} \in \mathcal{P}_{n+1}$. Setting $f(x) = \sum_{i=1}^\infty 2^{-i} \|f_i\|^{-1} \cdot f_i(x)$, where $\|\cdot\|$ denotes the supremum norm on $[a, b]$, $f(x)$ is clearly contained in $K_c[u_1, \ldots, u_n] \cap AC[a, b]$ and $D \begin{pmatrix} u_1 & \cdots & u_n & f \\ x_1 & \cdots & x_n & x_{n+1} \end{pmatrix} > 0$ for all $(x_1, \ldots, x_n, x_{n+1}) \in \mathcal{P}_{n+1}$. Hence, f is a function which we require.

(2) Let $\{u_i\}_{i=1}^n$ be an H_I-system in $C[a, b]$. By Proposition 2.3.1-(3),

$$D \begin{pmatrix} \sigma u_1 & \cdots & u_n \\ x_1 & \cdots & x_n \end{pmatrix} \geq 0 \quad \text{for all } (x_1, \ldots, x_n) \in \mathcal{P}_n,$$

where σ is a constant 1 or -1, and

$$\mathcal{P}'_n = \{(x_1, \ldots, x_n) \mid (x_1, \ldots, x_n) \in \mathcal{P}_n, D \begin{pmatrix} \sigma u_1 & \cdots & u_n \\ x_1 & \cdots & x_n \end{pmatrix} > 0\}$$

is dense in $\mathcal{P}_n$. Let $\{z_i\}_{i=1}^{\infty}$ be a countable subset of $\mathcal{P}'_n$ which is dense in $\mathcal{P}_n$. If each g_i, $i \in N$, is a function in $K_c[u_1, \ldots, u_n]$ which satisfies (3.4.1) and (3.4.3) for $z_i = (x_1, \ldots, x_n)$, without difficulty, we see that $g(x) = \Sigma_{i=1}^{\infty} 2^{-i} \|g_i\|^{-1} \cdot g_i(x)$ is a continuous function on $[a, b]$ which is adjoined to the H_I-space $[u_1, \ldots, u_n]$.

From Theorem 3.2.2 and Theorem 3.2.8, we have

Theorem 3.4.4. (1) *Every finite dimensional H_I-space of $L^1[a, b]$ has an adjoined function which belongs to $L^1[a, b]$.*

(2) *Every finite dimensional complete T_P-space of $C^1[a, b]$ has an adjoined function which is continuously differentiable on $[a, b]$.*

Proof. (1) Let $\{u_i\}_{i=1}^{n}$ be an H_I-system in $L^1[a, b]$. By Theorem 3.2.8, the system $\{v_0(x) = 1\} \cup \{v_i(x) = \int_a^x u_i(t)dt\}_{i=1}^{n}$ is a complete T_P-system in $AC[a, b]$. From Theorem 3.4.3, we have a function f in $AC[a, b]$ which is adjoined to the T_P-space $[v_0, v_1, \ldots, v_n]$. Again by Theorem 3.2.8, we easily see that f' belongs to $L^1[a, b]$ and is adjoined to the H_I-space $[u_1, \ldots, u_n]$.

(2) Since the existence of adjoined functions immediately follows in the case of one dimensional complete T_P-spaces, we only consider the cases of spaces whose dimensions are more than two. Let $\{u_i\}_{i=1}^{n}$ be a complete T_P-system in $C^1[a, b]$. Since u_1 has no zeros on $[a, b]$, set $w_i = u_i/u_1$, $i = 1, \ldots, n$. The system $\{w_1 = 1, \ldots, w_n\}$ is also a complete T_P-system. By Theorem 3.2.2, $\{w_i'\}_{i=2}^{n}$ is an H_I-system in $C[a, b]$. From Theorem 3.4.3, we obtain a function g in $C[a, b]$ which is adjoined to $[w_2', \ldots, w_n']$, that is, $\{w_2', \ldots, w_n', g\}$ is an H_I-system in $C[a, b]$. Again by Theorem 3.2.2, $\{w_1, \ldots, w_n, \int_a^x g(t)dt\}$ is a complete T_P-system in $C^1[a, b]$. Hence, $u_1(x) \int_a^x g(t)dt \in C^1[a, b]$ is adjoined to the complete T_P-space $[u_1, \ldots, u_n]$.

Let $\{u_i\}_{i=1}^{n}$ be a complete T_P-system such that each u_i, $i = 1, \ldots, n$, is analytic on $[a, b]$, i.e., each u_i, $i = 1, \ldots, n$, is analytic on an open interval containing $[a, b]$. We shall show an existence of adjoined anlytic functions to the complete T_P-space $[u_1, \ldots u_n]$.

If $\{u_i\}_{i=1}^{n}$ is a system of real-valued functions on (a, b) such that, for some $p, q \in \{0\} \cup N$, each $(x - a)^p u_i(x)(x - b)^q$, $i = 1, \ldots, n$, is analytic on $[a, b]$, then we call $\{u_i\}_{i=1}^{n}$ a system with *AE-property* (each u_i can be analytic at end points).

We state a variation of Theorem 3.2.2 for a complete T_P-system with AE-property.

Lemma 3.4.5. *Let* $\{1\} \cup \{u_i\}_{i=1}^n$ *be a system on* (a, b) *with AE-property. Then the following statements are equivalent:*

(1) $\{1\} \cup \{u_i\}_{i=1}^n$ *is a complete* T_P-*system on* (a, b).

(2) $\{u_i'\}_{i=1}^n$ *is a* WT_P-*system on* (a, b) *with AE-property and each* $u_i', i = 1, \ldots, n,$ *is represented as*

$$u_i'(x) = (x - a_1)^{2p_1} \cdots (x - a_k)^{2p_k} v_i(x)/(x - a)^p (x - b)^q, \qquad x \in (a, b),$$

$$\{a_j\}_{j=1}^k = V(\{u_i'\}_{i=1}^n) \cap (a, b), \ p, \ q \in \{0\} \cup N, \ p_i \in N, i = 1, \ldots, k,$$

where $\{v_i\}_{i=1}^n$ *is a comlete* T_P-*system on* (a, b) *such that* $v_i, i = 1, \ldots, n,$ *is analytic on* $[a, b]$.

Proof. Let $\{1\} \cup \{u_i\}_{i=1}^n$ be a complete T_P-system on (a, b). By AE-property of $\{u_i\}_{i=1}^n$, there are $r, s \in \{0\} \cup N$ and each $u_i, i = 1, \ldots, n,$ is represented as

$$u_i(x) = \frac{\omega_i(x)}{(x - a)^r (x - b)^s}, \qquad i = 1, \ldots, n, \quad x \in (a, b),$$

where each $\omega_i, i = 1, \ldots, n,$ is an analytic function on $[a, b]$. By Proposition 2.3.1 and Theorem 3.2.2, $\{u_i'\}_{i=1}^n$ is an H_I-system and hence a WT_P-system on any subinterval $[c, d]$ of (a, b). This means that $\{u_i'\}_{i=1}^n$ is a WT_P-system on (a, b). Since each $u_i', i = 1, \ldots, n,$ is represented as

$$u_i'(x) = \frac{z_i(x)}{(x - a)^{2r} (x - b)^{2s}}, \qquad x \in (a, b),$$

where $z_i(x) = \omega_i'(x)(x - a)^r (x - b)^s - \omega_i(x)\{r(x - a)^{r-1}(x - b)^s + s(x - b)^{s-1}(x - a)^r\}, i = 1, \ldots, n,$ $\{u_i'\}$ has AE-property and $\{z_i\}_{i=1}^n$ is a WT_P-system of analytic functions on $[a, b]$. Appling Theorem 3.3.6 to $\{z_i\}_{i=1}^n$, we obtain representations of $u_i', i = 1, \ldots, n$ which are desired.

Let $\{u_i'\}_{i=1}^n$ be a WT_P-system on (a, b) with AE-property. Since $V(\{u_i'\})$ is finite, by Proposition 2.3.1, $\{u_i'\}_{i=1}^n$ is an H_I-system on any subinterval $[c, d]$ of (a, b). From Theorem 3.2.2, it immediately follows that $\{1\} \cup \{u_i\}_{i=1}^n$ is a complete T_P-system on (a, b). This completes the proof.

Now we have

Theorem 3.4.6. *Let* $\{u_i\}_{i=1}^n$ *be a complete* T_P-*system of analytic functions on* $[a, b]$. *Then there exists an analytic function* u_{n+1} *on* $[a, b]$ *which is adjoined to the complete* T_P-*space* $[u_1, \ldots, u_n]$.

Proof. Let $\{u_i\}_{i=1}^n$ be a complete T_P-system on $[a, b]$. For the normalized complete T_P-system $\{1, v_1 = u_2/u_1, \ldots, v_{n-1} = u_n/u_1\}$, the system of derivatives $\{v_i'\}_{i=1}^{n-1}$ is a WT_P-system on $[a, b]$ by Proposition 2.3.1 and Theorem 3.2.2. From Theorem 3.3.6, we have

$$v_i'(x) = p_1(x)u_i^1(x), \qquad i = 1, \ldots, n-1,$$

where $p_1(x)$ is a polynomial which appeared in Theorem 3.3.6 and $\{u_i^1\}_{i=1}^{n-1}$ is a T_P-system on (a, b) of analytic functions on $[a, b]$. Without loss of generality, $\{u_i^1\}_{i=1}^{n-1}$ is a complete T_P-system on (a, b) by Corollary 2 in Zielke[39] or Theorem in Zalik[34].

Appling an analogous way stated above to the normalized complete T_P-system $\{1, v_1^1 = u_2^1/u_1^1, \ldots, v_{n-2}^1 = u_{n-1}^1/u_1^1\}$ with AE-property, from Lemma 3.4.5, we obtain a polynomial $p_2(x)$ and a complete T_P-system $\{u_i^2\}_{i=1}^{n-2}$ on (a, b) with AE-property. Hence, by repeating this procedure, we obtain a polynomial $p_{n-1}(x)$ and a complete T_P-system $\{u_1^{n-1}\}$ on (a, b) with AE-property.

For positive integers r and s, set $u_2^{n-1}(x) = ((x-a)^r - (b-a)^r)^s u_1^{n-1}(x)$, and by integration apply the above procedure in reverse to the complete T_P-system $\{u_1^{n-1}, u_2^{n-1}\}$ on (a, b) with AE-property. By Lemma 3.4.5, the system $\{1\} \cup \{v_i^1\}_{i=1}^{n-1}$, where

$$v_{n-1}^1(x) = \int_c^x p_2(x_1)u_1^2(x_1) \int_c^{x_1} \cdots \int_c^{x_{n-3}} p_{n-1}(x_{n-2})u_2^{n-1} dx_{n-2} \cdots dx_1 dx,$$

$$c, x \in (a, b),$$

is a complete T_P-system on (a, b). Since $p_1 u_1^1 v_{n-1}^1$ is analytic on $[a, b]$ for sufficiently large r, s, $\{v_i'\}_{i=1}^{n-1} \cup \{p_1 u_1^1 v_{n-1}^1\}$ is a WT_P-system on $[a, b]$ of analytic functions on $[a, b]$. Hence, by Proposition 2.3.1 and Theorem 3.2.2, $\{1\} \cup \{v_i\}_{i=1}^{n-1} \cup \{v_n = \int_c^x p_1(t)u_1^1(t)v_{n-1}^1(t)dt\}$ is a complete T_P-system on $[a, b]$ of analytic functions. Thus $\{u_i\}_{i=1}^n \cup \{u_1 v_n\}$ is a complete T_P-system which we require.

Corollary 3.4.7. *Let $\{u_i\}_{i=1}^n$ be a WT_P-system of analytic functions on $[a, b]$. Then there exists an analytic function u_{n+1} on $[a, b]$ which is adjoined to the WT_P-space $[u_1, \ldots, u_n]$.*

Proof. Since any $u \in [u_1, \ldots, u_n] - \{0\}$ does not vanish on a nondegenerate subinterval of $[a, b]$, by Proposition 2.3.1, $\{u_i\}_{i=1}^n$ is an H_I-system. Let $G = [1, \int_a^x u_1 dx, \ldots, \int_a^x u_n dx]$. By Theorem 3.2.2, G is a complete T_P-space. Hence, from Theorem 3.4.6, there exists an analytic function u_{n+1} on $[a, b]$ which is adjoined to G. Using

Theorem 3.2.2 again, we easily observe that u'_{n+1} is a analytic function on $[a, b]$ which is adjoined to $[u_1, \ldots, u_n]$.

3.5 Best L^1-Approximation by H_I-Systems

If $C[a, b]$ is endowed with the uniform norm, then by the results in Young[32] and Haar[3], the class of finite dimensional C-spaces coincides with that of T_P-spaces (see 2.3). When $C[a, b]$ is endowed with the L^1-norm, it is natural to study the following problem:

Find classes of finite dimensional C-spaces of $C[a, b]$.

To state precisely, we prepare

Notations and a Term. Let $L^1[a, b]$ be the space of all real-valued Lebesgue integrable functions on $[a, b]$. $\| \cdot \|_1$ denotes the L^1-norm on $L^1[a, b]$, i.e., $\|f\|_1 = \int_a^b |f| dx$ for all $f \in L^1[a, b]$. For a given subspace G of $L^1[a, b]$ and a given $f \in L^1[a, b]$, if there exists a best approximation to f from G with the L^1-norm, we call it a *best L^1-approximation* to f from G.

The first result of this problem is well known as Jackson's Theorem.

Theorem 3.5.1. (Jackson[5] and Krein[10]) *Every finite dimensional T_P-space in $(C[a, b], \| \cdot \|_1)$ is a C-space.*

Furthermore, in a general or concrete setting, many important results related to the problem have been obtained (e.g.[1, 4, 11, 17, 18, 19, 22, 26, 27, 31]). Among these results, we introduce a recent result by Strauss[31].

Definition. Let G be a finite dimensional subspace of $C[a, b]$ and let $G^* = \{f| f \in C[a, b], |f(x)| = |u(x)|$ for some $u \in G\}$. If for each $f \in G^* - \{0\}$, there exist a $u \in G - \{0\}$ such that

(1) $u(x) = 0$ a.e. on $Z(f) = \{x| x \in [a, b], f(x) = 0\}$,

(2) $u(x)\text{sgn}(f(x)) = |u(x)|$ for all $x \in [a, b] - Z(f)$,

then it is said that G satisfies the *A-property* or G is an *A-space*.

Theorem 3.5.2. (Strauss[31]) *Let G be a finite dimensional subspace of $(C[a, b], \| \cdot \|_1)$. If G is an A-space, it is a C-space.*

Remark 1. DeVore established a similar condition to the A-property for unique one-sided best L^1-approximation.

After this, generalizations of Theorem 3.5.2 and properties of A-spaces have been investigated in detail. The readers can obtain these in [12, 13, 14, 15, 16, 20, 21, 24, 28, 29, 30] and so on.

The main purpose of this section is to give a necessary and sufficient condition that H_T-spaces are C-spaces. To show this, we use the Hobby-Rice Theorem. It is well known that this theorem plays an important role in obtaining good L^1-approximations. The Hobby-Rice Theorem is the following:

Theorem 3.5.3. (Hobby and Rice[4]) *Let $\{u_i\}_{i=1}^n$ be linearly independent functions in $L^1[a,b]$. Then there exists points $a = a_0 < a_1 < \cdots < a_k < a_{k+1} = b$, $k \leq n$, such that*

$$\sum_{i=0}^{k} (-1)^i \int_{a_i}^{a_{i+1}} u_\ell(x)dx = 0, \qquad \ell = 1, 2, \ldots, n.$$

We first show that some results analogous to those in Micchelli[17] hold in the framework of H_T-systems.

When $\{u_i\}_{i=1}^n$ is a WT_P-system in $C[a,b]$, Micchelli[17] proved that $k = n$ for every set of points $\{a_i\}_{i=1}^k$ in Theorem 3.5.3, and gave a sufficient condition for the points to be uniquely determined.

Theorem 3.5.4. *Let $\{u_i\}_{i=1}^n$ be an H_T-system in $C[a,b]$. Then the points $(a <)$ $a_1 < \cdots < a_k(< b)$ in Theorem 3.5.3 are unique and $k = n$.*

Proof. Since every H_T-system in $C[a,b]$ is a WT_P-system, by Lemma 2 in Micchelli[17], we see that $k = n$ for every set of points $\{a_i\}_{i=1}^k$ in Theorem 3.5.3.

For any two sets of points $\{a_i\}_{i=1}^n$ and $\{b_i\}_{i=1}^n$ in Theorem 3.5.3, we have

$$\sum_{i=0}^{n} (-1)^i \int_{a_i}^{a_{i+1}} u_\ell(x)dx - \sum_{i=0}^{n} (-1)^i \int_{b_i}^{b_{i+1}} u_\ell(x)dx = 0, \qquad \ell = 1, \ldots, n.$$

If $\{a_i\}_{i=1}^n \neq \{b_i\}_{i=1}^n$, then there exist $k \leq n$ disjoint nondegenerate subintervals $\{I_j\}_{j=1}^k$ of $[a,b]$ such that

$$(3.5.1) \qquad 0 = \sum_{j=1}^{k} 2 \int_{I_j} \sigma_j u_\ell(x)dx, \qquad \ell = 1, \ldots, n,$$

where each $\sigma_j = 1$ or -1. But (3.5.1) contradicts the fact that $\{u_i\}_{i=1}^n$ is an H_T-system. Hence, we have $\{a_i\}_{i=1}^n = \{b_i\}_{i=1}^n$.

51

Remark 2. (1) Theorem 3.5.4 does not always follows from Theorem 2 in Micchelli[17]. Before showing an example, we recall that, for a system $\{u_i\}_{i=1}^n$, the convexity cone $K[u_1, \ldots, u_n]$ is the set of all real-valued functions f defined on (a, b) for which the determinants $D \begin{pmatrix} u_1 & \cdots & u_n & f \\ t_1 & \cdots & t_n & t_{n+1} \end{pmatrix}$ are nonnegative whenever $a < t_1 < \cdots < t_n < t_{n+1} < b$. Furthermore we denote by $K_c[u_1, \ldots, u_n]$ the set of all functions in $K[u_1, \ldots, u_n]$ which are continuous on $[a, b]$.

Let X be a nowhere dense and compact subset of (a, b) such that $\mu(X) > 0$, where μ is the Lebesgue measure. Then we consider a continuous function f such that $f(t) = \inf_{x \in X} |t - x|$. We easily observe that $\{f\}$ is an $H_{\mathcal{I}}$-system. Suppose that $g(t) \neq 0$ for some $g \in K_c[f]$ and some $t \in X$. Then some $h \in Span\{f, g\} - \{0\}$ which changes sign on (a, t) and (t, b) and $\{f, g\}$ is not a WT_P-system. Hence, we have $g(t) = 0$ for all $t \in X$ and $g \in K_c[f]$. Since $\|g - h\|_1 \geq \int_X |h| dx$ for all $g \in K_c[f]$ and all $h \in L^1[a, b]$, the smallest closed linear subspace of $(L^1[a, b], \| \cdot \|_1)$ containing $K_c[f]$ does not contain a one-dimensional T_P-space. Thus $\{f\}$ does not satisfy the condition of Theorem 2 in Micchelli[17].

(2) Suppose that $\{u_i\}_{i=1}^n$ is a WT_P-system such that the smallest closed linear subspace of $(C[a, b], \| \cdot \|_1)$ containing $K_c[u_1, \ldots, u_n]$ contains an $(n + 1)$-dimensional $H_{\mathcal{I}}$-space. Then, from an analogous method to the proof of Theorem 2 in Micchelli[17], the uniqueness of points in Theorem 3.5.3 can be derived.

(3) Other important results on the same problem as in Theorem 3.5.4 are shown in Sommer[26] and Kroó, Schmidt and Sommer[15].

Now we shall consider the uniqueness of the best L^1-approximation to a continuous function in the convexity cone from an $H_{\mathcal{I}}$-space.

Theorem 3.5.5. *Let $\{u_i\}_{i=1}^n$ be an $H_{\mathcal{I}}$-system in $C[a, b]$ and let $\{a_i\}_{i=1}^n$ be the unique set of points of Theorem 3.5.3 associated with this system. If $\{a_i\}_{i=1}^n \cap V(\{u_i\}_{i=1}^n) = \emptyset$, then the best L^1-approximation to f in $K_c[u_1, \ldots, u_n]$ from $[u_1, \ldots, u_n]$ is unique.*

Proof. By Proposition 2.3.1-(4), we observe that, for $a < t_1 < \cdots < t_n < b$ with $\{t_i\}_{i=1}^n \cap V(\{u_i\}_{i=1}^n) = \emptyset$,

$$(3.5.2) \qquad D \begin{pmatrix} u_1 & \cdots & u_n \\ t_1 & \cdots & t_n \end{pmatrix} \neq 0.$$

Hence, for any function $f \in K_c[u_1, \ldots, u_n]$, using (3.5.2), there is the unique function u_0 in $[u_1, \ldots, u_n]$ which interpolates f at $a_1, \ldots, a_n$. Though the rest of this proof,

52

that is, by showing that u_0 is the unique best L^1-approximation to f, similar to the proof in Theorem 1 in Micchelli[17], we can show the rest.

$f(x) - u_0(x)$ is expressed as a ratio of determinants

$$f(x) - u_0(x) = \frac{D\begin{pmatrix} u_1 & \cdots & u_n & f \\ a_1 & \cdots & a_n & x \end{pmatrix}}{D\begin{pmatrix} u_1 & \cdots & u_n \\ a_1 & \cdots & a_n \end{pmatrix}}.$$

By the above equation,

$$\sigma(-1)^{i+n}(f(x) - u_0(x)) \geq 0, \quad x \in (a_i, a_{i+1}), i = 0, \ldots, n,$$

where $\sigma = 1$ or -1, $a_0 = a$ and $a_{n+1} = b$. Hence, for any $u \in [u_1, \ldots, u_n]$, we obtain

$$\int_a^b |f(x) - u_0(x)| dx = \sigma \sum_{i=0}^n (-1)^{i+n} \int_{a_i}^{a_{i+1}} (f(x) - u_0(x)) dx$$

(3.5.3)
$$= \sigma \sum_{i=0}^n (-1)^{i+n} \int_{a_i}^{a_{i+1}} (f(x) - u(x)) dx$$

$$\leq \int_a^b |f(x) - u(x)| dx.$$

This means that u_0 is a best L^1-approximation to f. Suppose that $u \in G$ is a function such that $\|f - u_0\|_1 = \|f - u\|_1$. By (3.5.3), u must satisfy

$$\sigma(-1)^{i+n}(f(x) - u(x)) \geq 0, \quad x \in (a_i, a_{i+1}), i = 0, \ldots, n.$$

Hence, $f(a_i) = u(a_i)$, $i = 1, \ldots, n$. From (3.5.2), $u = u_0$ follows. This completes the proof.

Let G be a finite dimensional subspace of $C[a, b]$. By the result of Pinkus[20], subspaces of $C[a, b]$ which satisfy A-property are not spaces of H_I-type but generalized spline-like spaces. Here we shall consider another condition which ensures the uniqueness of best L^1-approximation from an H_I-space.

Notation and Definition. (1) Let $\{u_i\}_{i=1}^n$ be linearly independent functions in $C[a, b]$. For any point $x \in [a, b] - V(\{u_i\}_{i=1}^n)$, we can consider the closed subinterval $I_x = [a_x, b_x]$ of $[a, b]$, where

$$a_x = \sup\{a, \, y \mid y \in V(\{u_i\}_{i=1}^n) \text{ and } y < x\}$$

and

$$b_x = \inf\{b, \, z \mid z \in V(\{u_i\}_{i=1}^n) \text{ and } x < z\}.$$

53

Since $I_x = I_y$ or $I_x \cap I_y = \emptyset$ for any $x, y \in [a, b] - V(\{u_i\}_{i=1}^n)$, the set of subintervals $\{I_x\}_{x \in [a,b] - V(\{u_i\}_{i=1}^n)}$ essentially consist of at most countable disjoint subintervals $\{I_j\}_{j \in J}$, $J \subset N$. Then, putting

$$S[u_1, \ldots, u_n] = \{(\sigma_j)_{j \in J} \mid \sigma_j = 1 \text{ or } -1\},$$

we define, for each $\sigma \in S[u_1, \ldots, u_n]$ and each u_i, $i = 1, \ldots, n$, a continuous function

$$u_i^\sigma(t) = \begin{cases} \sigma_j u_i(t) & t \in I_j, j \in J \\ 0 & t \in V(\{u_i\}_{i=1}^n). \end{cases}$$

Let $\{a_i^\sigma\}_{i=1}^k$ be any set of points as in Theorem 3.5.3 for the system $\{u_i^\sigma\}_{i=1}^n$ for each $\sigma \in S[u_1, \ldots, u_n]$. Then $\{u_i\}_{i=1}^n$ is said to satisfy *condition B* if for every $\sigma \in S[u_1, \ldots, u_n]$, $k = n$ and $\{a_i^\sigma\}_{i=1}^k \cap V(\{u_i\}_{i=1}^n) = \emptyset$.

Remark 3. For example, T_P-systems and nonvanishing WT_P-systems satisfy condition B. In particular, A-spaces also satisfy condition B from the following facts. Let $\{u_i\}_{i=1}^n$ in $C[a, b]$ be a system with A-property.

(1) $[u_1^\sigma, \ldots, u_n^\sigma] = [u_1, \ldots, u_n]$ for any $\sigma \in S[u_1, \ldots, u_n]$ by Proposition 4.9 in Pinkus[20].

(2) By Theorem 1.1 in Kroó, Schmidt and Sommer[15], sets of points $\{a_i\}_{i=1}^k$ as in Theorem 3.5.3 are unique, $k = n$ and $\{a_i\}_{i=1}^n \cap V(\{u_i\}_{i=1}^n) = \emptyset$.

Using Theorem 23 and 24 in Cheney and Wulbert[1], we obtain

Lemma 3.5.6. *Every $f \in C[a, b]$ has a unique best L^1-approximation from a finite dimensional subspace G of $C[a, b]$ if and only if there does not exist a measurable function h and $g \in G - \{0\}$ such that (i) $|h(x)| = 1$, $x \in [a, b]$, (ii) $\int_0^1 h(x)u(x)dx = 0$ for all $u \in G$ and (iii) $h(x)|g(x)| \in C[a, b]$.*

Now we are in position to state

Theorem 3.5.7. *Let $\{u_i\}_{i=1}^n$ be an H_I-system in $C[a, b]$. Then any continuous function has a unique best L^1-approximation from $[u_1, \ldots, u_n]$ if and only if $\{u_i\}_{i=1}^n$ satisfies condition B.*

Proof. (Necessity) Suppose that, for the corresponding functions $\{u_i^\sigma\}_{i=1}^n$ to some $\sigma \in S[u_1, \ldots, u_n]$, there exists a set of points $\{a_i\}_{i=1}^k$ as in Theorem 3.5.3 such that $k < n$ or $\{a_i\}_{i=1}^k \cap V(\{u_i\}_{i=1}^n) \neq \emptyset$. We construct a continuous function f such that

$$Z(f) = \{a_i\}_{i=1}^k \cup V(\{u_i\}_{i=1}^n)$$

and

$$\text{sgn } f(t) = (-1)^i \cdot \sigma_j \quad t \in ([a,b] - Z(f)) \cap (a_i, a_{i+1}) \cap I_j, \quad i = 0, \ldots, k, j \in J.$$

On the other hand, noting that the number of points $|\{a_i\}_{i=1}^k - V(\{u_i\}_{i=1}^n)| < n$, we can find a g in $[u_1, \ldots, u_n] - \{0\}$ with $g(a_i) = 0$, $i = 1, \ldots, k$. Since $g(x) = 0$ for all $x \in V(\{u_i\}_{i=1}^n)$, we have $Z(g) \supset Z(f)$. Hence, if we set

$$h(x) = \begin{cases} 1 & x \in Z(f) \\ \text{sgn } f(x) & \text{otherwise,} \end{cases}$$

the functions $h(x)$ and $g(x)$ satisfy conditions (i) - (iii) in the Lemma 3.5.6, which means that $[u_1, \ldots, u_n]$ is not a unicity space in the L^1-norm.

(Sufficiency) Suppose that $[u_1, \ldots, u_n]$ is not a unicity space in the L^1-norm. Then, by Lemma 3.5.6, there exist functions $g(x) \in [u_1, \ldots, u_n] - \{0\}$ and a measurable function $h(x)$ which satisfy conditions (i) - (iii). Since g is a nonzero function in the $H_\mathcal{I}$-space $[u_1, \ldots, u_n]$, from the proof of Theorem 3.5.5, we have

$$(3.5.4) \qquad |Z(g) - (Z(\{u_i\}_{i=1}^n) \cup \{a, b\})| < n.$$

If we denote the set of points at which $h(x)$ is not continuous by D_h, condition (iii) means

$$(3.5.5) \qquad D_h \subset Z(g).$$

Combining condition (ii) with (3.5.4) and (3.5.5), we can take a $\sigma \in S[u_1, \ldots, u_n]$ and points $a = a_0 < a_1 < \cdots < a_k < a_{k+1} = b$, $k < n$ such that

$$(3.5.6) \qquad \sum_{i=0}^k (-1)^i \int_{a_i}^{a_{i+1}} u_\ell^\sigma(x) dx = 0, \qquad \ell = 1, \ldots, n.$$

But (3.5.6) contradicts the fact that $\{u_i\}_{i=1}^n$ satisfies condition B. This completes the proof.

Remark 4. The proof of necessity in Theorem 3.5.7 is valid for any system $\{u_i\}_{i=1}^n$ in $C[a,b]$ such that $[u_1, \ldots, u_n]$ is a unicity space.

Example 1. Let us consider the functions $f_1(t) = t^2$ and $f_2(t) = t^5$ in $C[-1,1]$. Clearly $[f_1, f_2]$ is a two-dimensional $H_\mathcal{I}$-space and the unique points in Theorem 3.5.3 are $-(1/2)^{1/3}$ and $(1/2)^{1/3}$. Since $Z(\{f_1, f_2\}) = \{0\}$, by Theorem 3.5.5, every function g in $K_c[f_1, f_2]$ has a unique best L^1-approximation from $[f_1, f_2]$. Furthermore, with $\sigma = 1$ on $[0,1]$ and -1 on $[-1,0]$, the corresponding functions f_1^σ and f_2^σ have the unique set of points $\{-(1/2)^{1/5}, (1/2)^{1/5}\}$ in Theorem 3.5.3. Hence, every function g in $C[-1,1]$ has a unique best L^1-approximation from $[f_1, f_2]$.

Example 2. Let f be a continuous function in $C[-3\pi/2, 3\pi/2]$ defined by

$$f(t) = \begin{cases} -\cos t & t \in [-3\pi/2, -\pi/2] \cup [\pi/2, 3\pi/2] \\ 2\cos t & t \in [-\pi/2, \pi/2]. \end{cases}$$

It is trivial that $[f]$ is a one dimensional $H_{\mathcal{I}}$-space with $Z(f) = \{\pm\pi/2, \pm 3\pi/2\}$ and the unique point in Theorem 3.5.3 is 0.

On the other hand, if we take a $\sigma = (\sigma_1, \sigma_2, \sigma_3) = (-1, 1, 1) \in S[f]$, where $I_1 = [-3\pi/2, -\pi/2]$, $I_2 = [-\pi/2, \pi/2]$, and $I_3 = [\pi/2, 3\pi/2]$, then the point $\pi/2$ is the unique point in Theorem 3.5.3 with the system $\{f^\sigma\}$. Combining these facts with Theorem 3.5.5 and 3.5.7, we see that the set of best L^1-approximations to some $f \in C[-3\pi/2, 3\pi/2]$ from $[f]$ is not singleton but that any function in $K_c[f]$ has a unique best L^1-approximation from $[f]$.

Let $\{u_i\}_{i=1}^n$ be an $H_{\mathcal{I}}$-system in $C[a, b]$ which satisfies condition B. The proof of sufficiency in Theorem 3.5.7 is valid for the systems $\{u_i^\sigma\}_{i=1}^n$, $\sigma \in S[u_1, \ldots, u_n]$. Hence, we can state

Corollary 3.5.8. *Let $\{u_i\}_{i=1}^n$ be an $H_{\mathcal{I}}$-system in $C[a, b]$ which satisfies condition B. Every $[u_1^\sigma, \ldots, u_n^\sigma]$, $\sigma \in S[u_1, \ldots, u_n]$, is a C-space in $(C[a, b], \|\cdot\|_1)$.*

Let $C^1[a, b]$ be the space of all real-valued continuously differentiable functions on $[a, b]$ and let $\|\cdot\|_v$ be the norm on $C^1[a, b]$ given by $\|f\|_v = |f(a)| + v(f)$, where $v(f)$ denotes the total variation of f on $[a, b]$. Then, from Theorem 3.5.7 and Theorem 3.2.2, we obtain

Corollary 3.5.9. *Let $\{1, u_1, \ldots, u_n\}$ be a complete T_P-system in $C^1[a, b]$. Then, in the function space $(C^1[a, b], \|\cdot\|_v)$, any function in $C^1[a, b]$ has a unique best approximation from $[1, u_1, \ldots, u_n]$ if and only if the system of derivatives $\{u_i'\}_{i=1}^n$ satisfies condition B.*

3.6 Problems

1. In 3.4, the existence of adjoined functions to complete T_P-spaces and $H_{\mathcal{I}}$-spaces of $AC[a, b]$ or $C[a, b]$ is shown, but we only answer a part of the following problem:

Let E be a function space such as $C[a, b]$, $C^1[a, b]$ and so on and let G be a finite dimensional Haar-like space of E. Are there adjoined functions in E to G? For example, does every complete $T_{\mathcal{P}}$ (resp.$WT_{\mathcal{P}}$)-space of $C^n[a, b]$ have an adjoined function in $C^n[a, b]$?

2. Let $\{u_i\}_{i=1}^n$ be an $H_{\mathcal{I}}$-system in $C[a, b]$. In 3.5, we show that the space $[u_1, \ldots, u_n]$ has a unique best L^1-approxiamtion to any $f \in C[a, b]$ if and only if $\{u_i\}_{i=1}^n$ satisfies condition B. Hence, we consider the following problem:

Find a class $\mathcal{C}$ of $WT_{\mathcal{P}}$-systems in $C[a, b]$ such that

(i) $\mathcal{C}$ strictly contains the class of all $H_{\mathcal{I}}$-systems in $C[a, b]$ and,

(ii) for each system $\{u_i\}_{i=1}^n \in \mathcal{C}$ with condition B, there exists a unique best L^1-approximation to any $f \in C[a, b]$ from $[u_1, \ldots, u_n]$.

For example, by Remark 3 in 3.5, we can consider a class $\mathcal{C}$ which contains a class of A-spaces.

Chapter 4 Approximation by Vector-Valued Monotone Increasing or Convex Functions

4.1 Introduction

Monotone increasing functions and convex functions on the real line play a fundamental part in real anlysis. Because these functions have some useful properties, which are highly geometric and intuitive, they can be easily generalized to an abstract setting. In the book by Roberts and Varberg[23], many properties about convex or generalized convex functions are stated in detail. Karlin and Studden[15] profoundly studied Tchebycheff systems, which are a generalization of monotone increasing or convex functions, related to the approximation theory and its statistical applications.

In approximation theory, some problems of best approximations by monotone increasing, convex and generalized convex functions have been treated in several function spaces with the usual norms.

We introduce two results in $C[a, b]$, which are related to problems in this chapter. Let $M[a, b]$ be the space of all real-valued bounded functions on $[a, b]$. Let $K_1[a, b]$ and $K_2[a, b]$ be the set of all monotone increasing ($=$ nondecreasing) functions in $F[a, b]$ and the set of all convex functions in $F[a, b]$, respectively.

Theorem 4.1.1.(Darst and Huotari[4;Lemma 1 and Theorem 8], Huotari and Zwick[11;(10)Theorem]) *In $L^1[a, b]$ with the L^1-norm, every $f \in C[a, b]$ has a unique best approximation $\tilde{f} \in C[a, b]$ from $K_i[a, b] \cap L^1[a, b]$, $i = 1, 2$.*

Theorem 4.1.2.(Darst and Sahab[2; Corollary 2], Zwick[33;Theorem 1]) *In $M[a, b]$ with the supremum norm, every $f \in C[a, b]$ has a best approximation $\tilde{f} \in C[a, b]$ from $K_i[a, b] \cap M[a, b]$, $i = 1, 2$.*

Many authors have studied further investigations of the above results and a lot of other important results of approximation by monotone increasing or convex functions. Ubhaya[27, 28] considered weighted uniform approximation by nondecreasing ($=$isotone) functions on a partially ordered set. Let f be a function which has a unique best L^p-approximation f_p for each $p \in (1, +\infty)$. Then, by considering $\lim_{p \to \infty} f_p$ (resp. $\lim_{p \to 1+0} f_p$), best L^∞ or L^1 approximations from nondecreasing, convex or n-convex functions are studied in Huotari et al.[6,7,9,12,13,21]. The uniqueness and description of best L^1 or L^p by nondecreasing functions are treated in [3,8,10,24,25].

Ubhaya et al.[29,30,31,32] obtained the existence and characterizations of best approximations by convex, quasi-convex or piecewise monotone functions. Swetits et al.[26] and Zwick[33,35,36,37] studied characterizations and other related results of best approximations by n-convex or generalized convex functions.

On the other hand, in subspaces of all R^m-valued functions on some set, Pinkus[22] has considered characterizations and the uniquesness of best approximations from finite dimensional subspaces.

In this chapter, we pay attention to a space of all vector-valued functions of bounded variation on a compact real interval and to the existence and uniqueness problems of best approximations by monotone increasing or convex functions.

Let $(E, \| \cdot \|)$ be an ordered Banach space over the real field. Let BV_E (resp. bv_E) be the space of all E-valued functions on a compact interval $[a, b] \subset R$ (resp. on $N_0 = N \cup \{\infty\}$) which are of bounded variation (resp. which are of bounded variation and continuous at ∞). When BV_E (resp. bv_E) is endowed with a norm $\|f\|_v = \|f(a)\| + v(f)$, $f \in BV_E$ (resp. $\|f\|_v = \|f(1)\| + v(f)$, $f \in bv_E$), where $v(f)$ is a total variation of f on $[a, b]$ (resp. on N_0), we are concerned with the existence and uniqueness of best approximations by monotone increasing or convex functions in BV_E (resp. bv_E).

In 4.2, basic properties of BV_E and bv_E are studied under the condition that E is a sequentially complete Hausdorff locally convex space. In 4.3, we show problems of the best approximations treated in this chapter. In 4.4, we study approximation by monotone increasing functions, and approximation by convex functions is investigated in 4.5. Problems related to this topic are given in 4.6.

4.2 On Spaces of Vector-Valued Functions of Bounded Variation

Let $E(\tau)$ be a sequentially complete Hausdorff locally convex space over the real or complex field. E' denotes the topological dual of $E(\tau)$. We write $\Gamma = \{p_\lambda \mid \lambda \in \Lambda\}$ for a system of saturated semi-norms on E generating the topology τ. Let $[a, b]$ be a linearly ordered interval. For simplicity, we write $F(\tau_p)$ for the product space $\Pi_{\alpha \in [a,b]} E_\alpha(\tau_\alpha)$ and write F' for the direct sum $\oplus_{\alpha \in [a,b]} E'_\alpha$, where $E_\alpha(\tau_\alpha) = E(\tau)$ and $E'_\alpha = E'$ for all $\alpha \in [a, b]$. The readers are referred to Köthe [20] for further explanation of the terminology used in this section.

59

Definition. Let $[a, b]$ and $E(\tau)$ be the linearly ordered interval and the locally convex space stated above. For an arbitrary E-valued function f defined on $[a, b]$, we set $V_\Delta(f, p_\lambda) = \sum_{i=0}^{n-1} p_\lambda(f(a_{i+1}) - f(a_i))$ for every $p_\lambda \in \Gamma$ and for every finite subset $\Delta = \{a_i, 0 \le i \le n, a = a_0 < a_1 < \cdots < a_n = b\}$ of $[a, b]$. Then, the extended real number $V(f, p_\lambda) = \sup_\Delta V_\Delta(f, p_\lambda)$ is called the *total variation* of f on $[a, b]$ with p_λ.

An E-valued function $f(x)$ is said to be of *bounded variation* if $V(f, p_\lambda) < +\infty$ for every $p_\lambda \in \Gamma$. The space of all E-valued functions on $[a, b]$ that are of bounded variation is denoted by BV_E.

Proposition 4.2.1. *BV_E is a dense subspace of $F(\tau_p)$.*

Proof. BV_E contains $\oplus_{\alpha \in [a,b]} E_\alpha$ which is dense in $F(\tau_p)$.

Notation 1. Using every semi-norm $p_\lambda \in \Gamma$, we define a semi-norm s_λ on BV_E by $s_\lambda(f) = p_\lambda(f(a)) + V(f, p_\lambda)$ for all $f \in BV_E$. The locally convex topology which is generated by the system of semi-norms $\{s_\lambda(\cdot)\}_{\lambda \in \Lambda}$ is denoted by τ_v.

If $\bar{\tau}_p$ is the induced topology on BV_E from $F(\tau_p)$, we obtain

Proposition 4.2.2. *τ_v is finer than $\bar{\tau}_p$.*

Proof. If, for each $\lambda \in \Lambda$, we take a τ_v-neighborhood $W = \{f \mid s_\lambda(f) \le 1\}$ of o, then $p_\lambda(f(\gamma)) \le p_\lambda(f(a)) + p_\lambda(f(\gamma) - f(a)) + p_\lambda(f(b) - f(\gamma)) \le s_\lambda(f) \le 1$ for all $f \in W$ and all $\gamma \in [a, b]$, which leads to our conclusion.

Since it is assumed that $E(\tau)$ is sequentially complete, we can show

Proposition 4.2.3. *$BV_E(\tau_v)$ is sequentially complete.*

Proof. Let $\{f_n\}_n$ be an arbitrary τ_v-Cauchy sequence. Since each sequence $(f_n(\alpha))_n$, $\alpha \in [a, b]$ is a Cauchy sequence in $E_\alpha(\tau_\alpha)$ by Proposition 4.2.2, we can define an E-valued function f on $[a, b]$ with $f(\alpha) = \lim_n f_n(\alpha)$, and we easily see that f belongs to BV_E and f_n converges to f in $BV_E(\tau_v)$.

Remark 1. (1) By Proposition 4.2.1, BV_E and F' form a dual pair.

(2) If the topological dual of $BV_E(\tau_v)$ is denoted by W_E, F' (which is the dual of $BV_E(\tau_p)$) is a subspace of W_E by Proposition 4.2.2.

Before giving the next proposition, we prepare some notations.

Notation 2. (1) For each $\lambda \in \Lambda$, let τ_λ be a locally convex topology on E

60

generated by the semi-norm p_λ (τ_λ is not necessarily Hausdorff.). Then, we denote the topological dual of $E(\tau_\lambda)$ by $E'(\lambda)$, which is a linear subspace of E'. We can give a well defined norm $\|\cdot\|_{\lambda'}$ on $E'(\lambda)$ such that $\|u\|_{\lambda'} = \min\{m \mid |<u,x>| \le m \cdot p_\lambda(x),\ x \in E\}$.

(2) For each $\lambda \in \Lambda$, we can define a norm $\|\cdot\|_\infty^\lambda$ on $F'_\lambda = \oplus_{\alpha \in [a,b]} E'_\alpha(\lambda)$, where $E'_\alpha(\lambda) = E'(\lambda)$ for each $\alpha \in [a,b]$, such that $\|v\|_\infty^\lambda = \sup_{1 \le j \le n} \|\Sigma_{i=1}^j v(a_i)\|_\lambda'$ for all $v = (v(\alpha))_\alpha \in F'_\lambda$, provided each $v(a_i), 1 \le i \le n$, is a nonzero element in $E'_{a_i}(\lambda)$ and $a \le a_1 < a_2 < \cdots < a_n \le b$.

Now we obtain

Proposition 4.2.4. *On the linear subspace $F'_\lambda = \oplus_{\alpha \in [a,b]} E'_\alpha(\lambda)$ of F', $B_m^\lambda = \{u \mid u = (u(\alpha))_\alpha \in F'_\lambda, \|u\|_\infty^\lambda \le m\}$ for each $m \in N$ and, for each $\lambda \in \Lambda$, B_m^λ is $\sigma(F', BV_E)$-bounded.*

Proof. For each $f \in BV_E$ and each $u \in B_1^\lambda$, we have

$$|<u,f>| = |\sum_{j=1}^n <\sum_{i=0}^{j-1} u(a_{n-i}), f(a_{n-j+1}) - f(a_{n-j})> + <\sum_{i=1}^n u(a_i), f(a)>|$$
$$\le s_\lambda(f),$$

where u vanishes except at $a_i, 1 \le i \le n$ and $a_0 = a$. Hence, this inequality shows the boundedness of B_1^λ.

Proposition 4.2.5. *The polar $B_m^{\lambda\circ}$ in BV_E of B_m^λ in Proposition 4.2.4 is $\{f \mid s_\lambda(f) \le 1/m, f \in BV_E\}$ for each $m \in N$ and for each $\lambda \in \Lambda$. Hence, each s_λ, $\lambda \in \Lambda$, is a $\beta(BV_E, F')$-continuous semi-norm.*

Proof. From the proof of Proposition 4.2.4, we have $s_\lambda(f) = V(f, p_\lambda) + p_\lambda(f(a)) = \sup\{|<u,f>| \mid \|u\|_\infty^\lambda \le 1, u \in F'_\lambda\}$. Hence, we obtain $B_1^{\lambda\circ} = \{f \mid s_\lambda(f) \le 1, f \in BV_E\}$. Similarly, the polar $B_m^{\lambda\circ}$ of B_m^λ in BV_E is shown by replacing 1 with $1/m$.

Proposition 4.2.6. *In BV_E, $\beta(BV_E, F')$-boundedness is identical with τ_v-boundedness.*

Proof. Since τ_v is finer than $\bar\tau_p$ on BV_E by Proposition 4.2.2, $\beta(BV_E, W_E)$ is finer than $\beta(BV_E, F')$. Further, $\beta(BV_E, F')$ is finer than τ_v by Proposition 4.2.5. On the other hand, τ_v-bounded and $\beta(BV_E, W_E)$-bounded subsets of BV_E are the same by

Proposition 4.2.3 and the proposition in Köthe [20; see Chap. 4, Section 20-11(3)]. Hence, $\beta(BV_E, F')$-boundedness is identical with τ_v-boundedness.

By Proposition 4.2.6, any $\beta(BV_E, F')$-bounded subset is contained in a subset of the form $B = \{f \mid s_\lambda(f) \leq M_\lambda, M_\lambda > 0$ for each $\lambda \in \Lambda, f \in BV_E\}$. Finally, we examine the closure in $F(\sigma(F, F'))$ of B.

Proposition 4.2.7. *The closure in $F(\sigma(F, F'))$ of every $\beta(BV_E, F')$-bounded subset is contained in BV_E.*

Proof. As is mentioned above, it is sufficient to show that each $\beta(BV_E, F')$-bounded subset $B = \{f \mid s_\lambda(f) \leq M_\lambda(> 0)$ for each $\lambda \in \Lambda, f \in BV_E\}$ is closed in $F(\sigma(F, F'))$. By the proofs of Proposition 4.2.4 and 4.2.5, every subset $B_\lambda = \{f \mid s_\lambda(f) \leq M_\lambda, M_\lambda > 0, f \in F\}$, $\lambda \in \Lambda$, is $\sigma(F, F')$-closed. Since $B = \cap_{\lambda \in \Lambda} B_\lambda \subset BV_E$, we can easily verify the above fact.

If $E(\tau)$ is a locally convex space such that each strongly bounded subset is relatively compact in $E(\tau)$, then we call it a *β-semi-Montel space*. If a β-semi-Montel space is infra-barrelled, then it is called a *β-Montel space*. (see Definition in Kitahara[16])

We consider the following question: For what locally convex spaces $E(\tau)$ is every bounded subset of $BV_E(\sigma(BV_E, F'))$ such that, for each semi-norm $p_\lambda \in \Gamma$, the values of total variation are uniformly bounded (i.e., every τ_v-bounded subset) relatively $\sigma(BV_E, F')$-compact ?

Since a subset B of $BV_E(\sigma(BV_E, F'))$ is the bounded subset mentioned above if and only if B is strongly bounded by Proposition 4.2.6, the problem we consider is transformed as follows: under what locally convex space $E(\tau)$ is every strongly bounded subset of $BV_E(\sigma(BV_E, F'))$ relatively compact or is $BV_E(\sigma(BV_E, F'))$ a β-semi-Montel space?

Now we can state

Theorem 4.2.8. *$BV_E(\sigma(BV_E, F'))$ is β-semi-Montel if and only if $E(\tau)$ is semi-reflexive. In particular, $BV_E(\sigma(BV_E, F'))$ is semi-Montel if and only if $E(\tau)$ is semi-reflexive and $[a, b]$ is a finite set.*

Proof. If $BV_E(\sigma(BV_E, F'))$ is β-semi-Montel, the closed subspace $E_0 = \{f \mid f(a) = x, x \in E, f(y) = 0$ for all $y \in (a, b]\}$ of $BV_E(\sigma(BV_E, F'))$ is β-semi-Montel by Proposition 5 in Kitahara[16]. Furthermore, $E_0(\sigma(BV_E, F'))$ is linearly homeomorphic

62

to $E(\sigma(E, E'))$ (see Chap.4, Section 22-5-(3) in Köthe[20]). Since $E(\tau)$ is sequentially complete, $\sigma(E, E')$-boundedness is identical with $\beta(E, E')$-boundedness by the same argument used in the proof of Proposition 4.2.6. Thus, an arbitrary closed $\sigma(E, E')$-bounded subset is $\sigma(E, E')$-compact.

Conversely, suppose that $E(\tau)$ is semi-reflexive and that $F(\sigma(F, F'))$ is semi-Montel. Since an arbitrary closed $\beta(BV_E, F')$-bounded subset in $BV_E(\sigma(BV_E, F'))$ is $\sigma(F, F')$-bounded and closed in $F(\sigma(F, F'))$ by Proposition 4.2.7, it is $\sigma(BV_E, F')$-compact. Hence, $BV_E(\sigma(BV_E, F'))$ is β-semi-Montel.

Suppose that $BV_E(\sigma(BV_E, F'))$ is semi-Montel and that $[a, b]$ is an infinite set. Then, we can easily find a Cauchy sequence which converges to a function of which total variation is not finite, which leads to a contradiction. Finally, if $E(\tau)$ is semi-reflexive and $[a, b]$ is a finite set, $BV_E(\sigma(BV_E, F'))$ is semi-Montel from $F = BV_E$.

Corollary 4.2.9. *If $E(\tau)$ is a Banach space, $BV_E(\sigma(BV_E, F'))$ is β-semi-Montel if and only if $E(\tau)$ is reflexive.*

From sequence spaces, we set $bv = \{f \mid f \in K^N, K \text{ is } R \text{ or } C, \lim_{n \to \infty} \sum_{i=1}^{n} |f(i+1) - f(i)| \text{ exists. }\}$ (see §2 in Chapter 2 in Kamthan and Gupta[14]).

Corollary 4.2.10. *The sequence space $bv(\tau_0)$ endowed with the topology of simple convergence is β-Montel.*

Proof. As a linearly ordered interval, we consider the positive integers and infinity Z_∞^+ with the usual order. By Theorem 4.2.8, $BV_K[Z_\infty^+](\tau_s)$ is β-semi-Montel, where K is the real or complex field and τ_s is the topology of simple convergence. Then, $bv_K(\bar{\tau}_s) = \{f \mid f \in BV_K[Z_\infty^+](\tau_s), f(\infty) = \lim_{n \to \infty} f(n)\}$ is linearly homeomorphic to $bv(\tau_0)$ and is a closed subspace of $BV_K[Z_\infty^+](\tau_s)$. Thus $bv(\tau_0)$ is β-semi-Montel by Proposition 5 in Kitahara[16]. Clearly, since each strongly bounded subset in the weak dual of $bv(\tau_0)$ is finite dimensional, $bv(\tau_0)$ is infra-barrelled .

Remark 2. (1) Theorem 4.2.8 is an extension of the classical result by Helly[5], which is called Helly's selection principle.

(2) Properties of β-semi-Montel or β-Montel spaces or related facts to these spaces are studied in Kitahara[16,17,18].

4.3 Problems on Best Approximations by Monotone Increasing or Convex Functions

First, we make some preparations for the problems treated here. Let E be a real Banach space with a norm $\|\cdot\|$ and let C be a proper closed convex cone of E with vertex o. Throughout this chapter, C is simply called a *cone*. By a cone C of E, we induce an order relation such that $x \leq y$ if and only if $y - x \in C$ and if $x \leq y$ and $x \neq y$, then $x < y$. $F_E[a, b]$ (resp. $F_E(N_0)$) denotes the space of all E-valued functions on a compact interval $[a, b] \subset R$ (resp. on $N_0 = N \cup \{\infty\}$ with the usual order). For an f of $F_E[a, b]$ (resp. $F_E(N_0)$), we consider the total variation $v(f)$ of f, that is,

$$v(f) = \sup_{\{x_0, \ldots, x_{n+1}\} \in \mathcal{D}} \sum_{i=0}^{n} \|f(x_{i+1}) - f(x_i)\|,$$

where $\mathcal{D}$ is a family of all finite subsets $\{x_0, \cdots, x_{n+1}\}$ of $[a, b]$ (resp. N_0) with $a = x_0 < \cdots < x_{n+1} = b$ (resp $1 = x_0 < \cdots < x_{n+1} = \infty$). If the total variation of f of $F_E[a, b]$ (resp. $F_E(N_0)$) is finite, f is said to be of bounded variation. The space of all E-valued functions of bounded variation is denoted by BV_E (resp. $bv_E = \{f \mid f$ is an E-valued function of bounded variation on N_0 and $f(\infty) = \lim_{n \to \infty} f(n)\}$) and we endow a norm on BV_E (resp. bv_E) such that $\|f\|_v = \|f(a)\| + v(f)$, $f \in BV_E$ (resp. $\|f\|_v = \|f(1)\| + v(f)$, $f \in bv_E$). As subsets of BV_E and bv_E, we set the followings.

$K_1 := \{f \mid f \in BV_E, f(x) \leq f(y) \text{ for } x \leq y\}$

$K_2 := \{f \mid f \in BV_E, \dfrac{f(x_2) - f(x_1)}{x_2 - x_1} \leq \dfrac{f(x_3) - f(x_2)}{x_3 - x_2} \text{ for } x_1 < x_2 < x_3\}$

$k_1 := \{f \mid f \in bv_E, f(i) \leq f(j) \text{ for } i \leq j\}$

$k_2 := \{f \mid f \in bv_E, f(i + 1) - f(i) \leq f(i + 2) - f(i + 1) \text{ for } i \in N\}$

Any functions (resp. sequences) in K_1 (resp. k_1) are called monotone increasing functions (resp. monotone increasing sequences), and any functions (resp. sequences) which belong to K_1 (resp. k_1) are called convex functions (resp. convex sequences).

For a given f in $BV_E - K_i$ (resp. $bv_E - k_i$), if there is an $\tilde{f}$ in K_i (resp. k_i) such that $\|f - \tilde{f}\|_v = \inf\{\|f - g\|_v \mid g \in K_i \text{ (resp. } g \in k_i)\}$, then $\tilde{f}$ is called a best approximation to f by K_i (resp. k_i). The set of all best approximations to f by K_i (resp. k_i) is denoted by $P_i(f)$ (resp. $p_i(f)$). Now, we consider the following problems: for an f in $BV_E - K_i$ or $bv_E - k_i$, $i = 1, 2$,

(P.1) does there exist a best approximation to f by K_i or k_i ?

(P.2) if f is continuous, does $P_i(f)$ contain continuous functions ?

(P.3) if f is continuously differentiable, does $P_i(f)$ contain continuously differentiable functions ?

Before answering (P.1), we prove

Lemma 4.3.1. *Let $E(\tau)$ be an Banach space and let F be the space stated in 4.2. Then*

(1) K_i, $i = 1, 2$, *are closed in* $BV_E(\sigma(BV_E, F))$.

(2) k_i, $i = 1, 2$, *are closed in* $bv_E(\sigma(bv_E, F))$.

Proof. We shall only prove the closedness of K_2 in $BV_E(\sigma(BV_E, F))$. As for K_1, k_1 and k_2, analogous methods to the case of K_2 can be applied. For any $f \in BV_E - K_2$, there are three points $(a \le)x_1 < x_2 < x_3(\le b)$ such that

$$\frac{f(x_3) - f(x_2)}{x_3 - x_2} - \frac{f(x_2) - f(x_1)}{x_2 - x_1} \notin C.$$

Since C is a closed convex set in $E(\tau)$, it is closed in $E(\sigma(E, E'))$. There exists a $\sigma(E, E')$-neighbourhood $B := \{x \mid x \in E, \mid < u_i, x > \mid \le 1, u_i \in E', i = 1, \ldots, n\}$ of o ($\mid <, > \mid$ denotes the bilinear form.) such that

$$(4.3.1) \qquad (\frac{f(x_3) - f(x_2)}{x_3 - x_2} - \frac{f(x_2) - f(x_1)}{x_2 - x_1} + B) \cap C = \emptyset.$$

If we set $\delta = \min\{(x_3 - x_2)/4, (x_2 - x_1)/4\}$ and consider a neighbourhood U of f in $BV_E(\sigma(BV_E, F))$ such that $U = f + \{h \mid h \in BV_E, \mid < u_i, h(x_j) > \mid \le \delta, i = 1, \ldots, n, j = 1, 2, 3\}$, then, by (4.3.1), the intersection of U and K_2 is empty. Hence, K_2 is closed in $BV_E(\sigma(BV_E, F))$.

First, we state a general result of the existence of best approximations in $(BV_E, \|\cdot\|_v)$

Theorem 4.3.2. *Let $(E, \|\cdot\|)$ be a real reflexive Banach space and let B be any closed subset in $BV_E(\sigma(BV_E, F))$. Then, for any $f \in (BV_E, \|\cdot\|_v)$, there exist best approximations by B.*

Proof. For a given f in $BV_E - B$, set $d = \inf\{\|f - g\|_v \mid g \in B\}$ and $A_\epsilon = B \cap \{h \mid \|f - h\|_v \le d + \epsilon\}$, $\epsilon > 0$. Since $(E, \|\cdot\|)$ is reflexive, by Theorem 4.2.8 and Lemma 4.3.1, each A_ϵ is compact in $BV_E(\sigma(BV_E, F))$. Noting that the family $\{A_\epsilon\}_{\epsilon > 0}$ has finite intersection property, the subset of B, $A = \cap_{\epsilon > 0} A_\epsilon$ is not empty. Hence, we easily see that each function in A is a best approximation to f by B.

As for (P.1), we can answer affirmatively under the following condition.

Corollary 4.3.3. *Let $(E, \|\cdot\|)$ be a real reflexive Banach space with a cone C. For a given f in $BV_E - K_i$ (resp. $bv_E - k_i$), $i = 1, 2$, there exists a best approximation $\tilde{f}$ to f by K_i.*

Remark. (1) Lemma 4.3.1 holds for arbitrary locally convex spaces.

(2) Since $v(g(x) + z) = v(g(x))$ for $g \in BV_E$ and any element $z \in E$, $\tilde{f}(a) = f(a)$ for every best approximation $\tilde{f}$ to f by K_i, $i = 1, 2$.

For convenience, we answer these problems in the case of approximation by monotone increasing functions and in the case of approximation by convex functions, respectively.

4.4 Approximation by Monotone Increasing Functions

First we prepare the following lemmas.

Lemma 4.4.1.(Proposition[3.1] in Chap. 1 in Barbu and Precupanu[1]) *If f belongs to BV_E, $f(x_0 + 0) = \lim_{x \to x_0, x > x_0} f(x)$ and $f(x_0 - 0) = \lim_{x \to x_0, x < x_0} f(x)$ exist in the norm topology at all $x_0 \in [a, b)$ and $x_0 \in (a, b]$, respectively.*

If $(E, \|\cdot\|)$ is a real reflexive strictly convex Banach space with a cone C, there exists a unique nearest point $\tilde{x}$ in C to any $x \in E$, i.e., $\|x - \tilde{x}\| = \inf_{y \in C} \|x - y\|$. Hence, we can define the projective function $P_C : E \to C$ by $P_C x = \tilde{x}$. This function is called the *projection* into C.

Lemma 4.4.2. *Let $(E, \|\cdot\|)$ be a real reflexive strictly convex Banach space with a cone C and let f be a continuous function in BV_E. Then the E-valued function $P_C \circ f(x)$ is continuous from $[a, b]$ to E with the weak topology τ_s.*

Proof. It is sufficient to show the map $P_C : (E, \|\cdot\|) \to (E, \tau_s)$ is continuous. Suppose that P_C is not continuous at $x_0 \in E$. Let $\{x_n\}$ be a converging sequence to x_0 in $(E, \|\cdot\|)$ and $\{P_C(x_n)\}$ does not converge to $P_C(x_0)$ in (E, τ_s). Since $\|P_C(x_n)\| \le \|P_C(x_n) - x_n\| + \|x_n\|$, $n \in N$, $\{P_C(x_n)\}$ is bounded. By reflexivity of $(E, \|\cdot\|)$, a subsequence $\{P_C(x_{n_k})\}$ of $\{P_C(x_n)\}$ converges to $z(\neq P_C(x_0))$ in the weak topology. Hence, we have

$$\|x_0 - z\| \le \lim_{k \to \infty} \|x_{n_k} - P_C(x_{n_k})\| = \|x_0 - P_C(x_0)\|.$$

But this contradicts the fact that $P_C(x_0)$ is the unique nearest point to x_0 in C.

Lemma 4.4.3.(Theorem 3.3 in Chap. 1 in Barbu and Precupanu[1]) *Let* $(E, \|\cdot\|)$ *be a real reflexive Banach space and let f be a function in BV_E. Then, the weak derivative f'_w of f exists almost everywhere on (a, b). Furthermore f'_w is integrable on $[a, b]$ and* $\int_a^b \|f'_w(t)\| dt \leq v(f)$.

Now we are in position to state

Theorem 4.4.4. *Let* $(E, \|\cdot\|)$ *be a real reflexive strictly convex Banach space with a cone C and let f be a function in $BV_E - K_1$. Then*

(1) *if f is continuous, $P_1(f)$ consists of continuous functions,*

(2) *if f is continuously differentiable, f has a unique best approximation $\tilde{f}$, which is expressed as* $\tilde{f}(x) = \int_a^x P_C \circ f'(t) dt$.

Proof. (1) Let $\tilde{f}$ be a best approximation to f by K_1 and suppose that $\tilde{f}$ is not continuous at $x_0 \in [a, b]$. For convenience, assume that x_0 lies in (a, b). By Lemma 4.4.1, $\tilde{f}(x_0 + 0)$ and $\tilde{f}(x_0 - 0)$ exist and $\|\tilde{f}(x_0 + 0) - \tilde{f}(x_0 - 0)\| > 0$. Hence, we set another nondecreasing function h such that

$$h(x) = \begin{cases} \tilde{f}(x) & x \in [a, x_0) \\ \tilde{f}(x_0 - 0) & x = x_0 \\ \tilde{f}(x) - (\tilde{f}(x_0 + 0) - \tilde{f}(x_0 - 0)) & x \in (x_0, b]. \end{cases}$$

Since $v(h) < v(\tilde{f})$, we reach a contradiction.

(2) Let f be a given continuously differentiable function in $BV_E - K_1$ and let h be any function in K_1. By Lemma 4.4.3, h is weakly differentiable a.e. on (a, b) and

$$(4.4.1) \qquad \int_a^b \|f' - h'_w\| dt \leq v(f - h).$$

Since h is increasing, $h'_w(x)$ belongs to C a.e. on (a, b). If h'_w belongs to C at x_0, since

$$\|f'(x_0) - h'_w(x_0)\| \geq \inf_{z \in C} \|f'(x_0) - z\| = \|f'(x_0) - P_C \circ f'(x_0)\|$$

and $P_C \circ f'$ is integrable on $[a, b]$ by Lemma 4.4.2, we have

$$(4.4.2) \qquad \int_a^b \|f' - P_C \circ f'\| dt \leq \int_a^b \|f' - h'_w\| dt.$$

In this inequality, equality holds only when $P_C \circ f' = h'_w$ a.e. on (a, b). Hence, if we put $\tilde{f} = f(a) + \int_a^x P_C \circ f'(t) dt$, by (4.4.1) and (4.4.2), $\tilde{f}$ is a unique best approximation to f by K_1.

Corollary 4.4.5. *Let* $(E, \|\cdot\|)$ *be a uniformly convex real Banach space with a cone* C *and let* f *be a function in* $BV_E - K_1$. *If* f *is continuously differentiable on* $[a, b]$, f *has a unique best approximation, which is continuously differentiable on* $[a, b]$.

Proof. Since P_C is a continuous function from $(E, \|\cdot\|)$ to $(E, \|\cdot\|)$ by uniform convexity of $(E, \|\cdot\|)$, the conclusion follows.

From the proof of Theorem 4.4.4, we obtain

Proposition 4.4.6. *Let* $(E, \|\cdot\|)$ *be a real reflexive strictly convex Banach space with a cone* C *and let* f *be a sequence in* $bv_E - k_1$. *Then* f *has a unique best approximation* $\tilde{f}$, *which is expressed as*

$$\tilde{f}(i) = \begin{cases} \tilde{f}(1) & i = 1 \\ \tilde{f}(1) + \sum_{j=1}^{i-1} P_C(f(j+1) - f(j)) & i \geq 2. \end{cases}$$

Proof. By Corollary 4.3.3, $p_1(f)$ is not empty. For any $g \in p_1(f)$, we see that

$$v(f - g) = \|f(1) - g(1)\| + \sum_{i=1}^{\infty} \|f(i+1) - f(i) - (g(i+1) - g(i))\|.$$

Furthermore, we have

$$\|f(i+1) - f(i) - P_C(f(i+1) - f(i))\| \leq \|f(i+1) - f(i) - (g(i+1) - g(i))\|, \quad i \in N.$$

Since clearly $\tilde{f} \in k_1$ and $\tilde{f} \in p_1(f)$, in the above inequality, the equlities must hold for each $i \in N$. This implies that f has a unique best approximation by k_1.

Example. Let $(E, \|\cdot\|)$ be a real line $(R, |\cdot|)$ with $C = \{x | x \geq 0\}$. For a given f of $BV_R - K_1$, let $p_f(x)$ and $n_f(x)$ be the positive and negative variation of f, respectively. Then $f(x)$ is expressed as $f(x) = f(a) + p_f(x) - n_f(x)$, $x \in [a, b]$. For any $g \in K_1$, since it holds that

(4.4.3) $\qquad n_{f-g}(y) - n_{f-g}(x) \geq n_f(y) - n_f(x) \qquad a \leq x \leq y \leq b$,

we have

$$\|f - g\|_v = |(f - g)(a)| + p_{f-g}(b) + n_{f-g}(b) \geq n_f(b).$$

By this inequality,

(4.4.4) $\qquad \inf\{\|f - g\|_v \mid g \in K_1\} = n_f(b)$

and $g_1(x) = f(a) + p_f(x)$ is a best approximation to f by K_1. Let g_2 be any best approximation to f by K_1. From the Remark in 4.3, (4.4.3) and (4.4.4),

$$g_2(a) = f(a), \quad p_{f-g_2}(x) = 0, \quad n_{f-g_2}(x) = n_f(x) \qquad x \in [a, b]$$

follow, which means $g_1 \equiv g_2$. Hence, $f(a) + p_f(x)$ is the unique best approximation to f by K_1.

4.5 Approximation by Convex Functions

First, we give a lemma with Lipschitz continuity of convex functions.

Lemma 4.5.1. *Let $(E, \| \cdot \|)$ be a real Banach space with a cone C. If each $\lfloor o, u \rfloor := \{x \mid o \leq x \leq u, x \in E\}$, $u \in C$, is bounded, then every convex function in K_2 is Lipschitz continuous on any closed subinterval $[c, d] \subset (a, b)$.*

Proof. Let f be a convex function on $[a, b]$. For a closed subinterval $[c, d] \subset (a, b)$ and for any points $x_1, x_2 \in [c, d]$ with $x_1 < x_2$, we obtain

$$u = \frac{f(c) - f(a)}{c - a} \leq \frac{f(x_2) - f(x_1)}{x_2 - x_1} \leq \frac{f(b) - f(d)}{b - d} = v$$

and

$$o \leq \frac{f(x_2) - f(x_1)}{x_2 - x_1} - u \leq v - u.$$

From the condition of this lemma, we have

$$\|(f(x_2) - f(x_1))/(x_2 - x_1) - u\| \leq K,$$

where K is a constant, and

$$\|f(x_2) - f(x_1)\| \leq (\|u\| + K)|x_2 - x_1|.$$

Remark 1. Under the condition of Lemma 4.5.1, every $f \in K_2$ is absolutely continuous on $[a, b]$ and if $(E, \| \cdot \|)$ is reflexive, then f is expressed as

$$f(x) = f(a) + \int_a^x \varphi(t)dt \qquad x \in [a, b],$$

where $\varphi(t)$ is increasing a.e. on (a, b) and integrable on $[a, b]$.

As for (P.2) and (P.3), we can state the following theorem under some additional conditions.

Theorem 4.5.2. *Let $(E, \| \cdot \|)$ be a real reflexive Banach space with a cone C and let each $\lfloor o, u \rfloor$, $u \in C$, be bounded.*

(1) *For a given f in $BV_E - K_2$, if f is continuous on $[a, b]$, $P_2(f)$ consists of continuous functions on $[a, b]$.*

Furthermore, suppose that $(E, \| \cdot \|)$ is uniformly convex and

(C.1) $\|u\| \le \|v\|$ *for all $o \le u \le v$.*

(2) *For a given f of $BV_E - K_2$, if f is continuously differentiable on $[a, b]$, f has a unique best approximation by K_2, which is continuously differentiable on (a, b).*

Proof. (1) We shall prove this by an analogous method to the proof of Theorem 4.4.4. Let $\tilde{f}$ be any best approximation to f in $BV_E - K_2$ by K_2. By Lemma 4.5.1, $\tilde{f}$ is continuous on (a, b). Since f is continuous on $[a, b]$, $\tilde{f}$ must be continuous at a and b. It is because, if $f(a) = \tilde{f}(a) \ne \tilde{f}(a + 0)$, then, by putting

$$h(x) = \begin{cases} f(a) & x = a \\ \tilde{f}(x) + f(a) - \tilde{f}(a + 0) & \text{otherwise,} \end{cases}$$

we have $v(f - \tilde{f}) > v(f - h)$, which contradicts the fact that $\tilde{f}$ is a best approximation to f by K_2. Similarly way, the continuity of f at b is shown.

(2) Let $\tilde{f}$ be any best approximation to f by K_2. Since $\tilde{f}$ belongs to K_2, by Remark 1, $\tilde{f}$ is absolutely continuous and is expressed as

$$\tilde{f}(x) = f(a) + \int_a^x \varphi(t) dt \qquad x \in [a, b],$$

where $\varphi(t)$ is increasing a.e. on (a, b) and integrable on $[a, b]$. Hence, without loss of generality, we assume that φ is increasing on (a, b).

Suppose to the contrary that $\varphi(t)$ is not continuous on (a, b), i.e., by (C.1), there exist a point $c \in (a, b)$ and a fixed number $\epsilon_0 > 0$ satisfying

(4.5.1) $\|\varphi(x) - \varphi(y)\| \ge \epsilon_0$ for all $x \in (a, c)$ and $y \in (c, d)$.

For any $\alpha \in (a, c)$ and $\beta \in (c, b)$, set

(4.5.2) $L_\alpha := \bigcup_{\alpha \le x < c} \lfloor \varphi(\alpha), \varphi(x) \rfloor, \quad R_\beta := \bigcup_{c < y \le \beta} \lfloor \varphi(y), \varphi(\beta) \rfloor.$

Since φ is increasing, L_α and R_β are convex and the closure $\bar{L}_\alpha$ and $\bar{R}_\beta$ of L_α and R_β are also convex. By (4.5.1) and (4.5.2), for all $z_1 \in \bar{L}_\alpha$ and all $z_2 \in \bar{R}_\beta$,

(4.5.3) $\|z_1 - z_2\| \ge \epsilon_0$

(4.5.4) $\varphi(\alpha) \le z_1 < z_2 \le \varphi(\beta).$

70

Let $z(\alpha)$ and $z(\beta)$ be the unique nearest points to $f'(c)$ in $\bar{L}_\alpha$ and $\bar{R}_\beta$ respectively and set $\ell_\alpha := \| z(\alpha) - f'(c)\|$ and $r_\beta := \|z(\beta) - f'(c)\|$. Since, from (C.1), we have

$$\max\{\|z(\alpha) - f'(c)\|, \|z(\beta) - f'(c)\|\}$$
$$\leq \max\{\|z(\alpha)\|, \|z(\beta)\|\} + \|f'(c)\|$$
$$\leq \|\varphi(\alpha)\| + \|\varphi(\beta) - \varphi(\alpha)\| + \|f'(c)\|,$$

ℓ_α and r_β increasingly converge to constants ℓ_0 and r_0 as α and β converge to c respectively.

Without loss of generality, let us consider the case $r_0 \geq \ell_0$ and $r_0 \neq 0$. Let α_0, β_0 be fixed numbers with $a < \alpha_0 < c$ and $c < \beta_0 < b$. For any ϵ_1, ϵ_2 with $0 < \epsilon_1 < c - \alpha_0$ and $0 < \epsilon_2 < \beta_0 - c$, we obtain

$$\int_{c-\epsilon_1}^{c} \|f'(t) - \varphi(t)\|dt = \int_{c-\epsilon_1}^{c} \|f'(t) - \frac{z(c-\epsilon_1) + z(c+\epsilon_2)}{2}$$
$$+ \frac{z(c-\epsilon_1) + z(c+\epsilon_2)}{2} - \varphi(t)\|dt$$
$$\leq \int_{c-\epsilon_1}^{c} \|f'(t) - \frac{z(c-\epsilon_1) + z(c+\epsilon_2)}{2}\|dt$$
$$+ \int_{c-\epsilon_1}^{c} \|\frac{z(c-\epsilon_1) + z(c+\epsilon_2)}{2} - \varphi(t)\|dt$$

and, from (C.1) and (4.5.4),

$$|\int_{c-\epsilon_1}^{c} \|f'(t) - \varphi(t)\|dt - \int_{c-\epsilon_1}^{c} \|f'(t) - \frac{z(c-\epsilon_1) + z(c+\epsilon_2)}{2}\| \, dt \,|$$
$$\leq \int_{c-\epsilon_1}^{c} \|\frac{z(c-\epsilon_1) + z(c+\epsilon_2)}{2} - \varphi(t)\|dt$$

(4.5.5)
$$\leq \int_{c-\epsilon_1}^{c} 2(\|\varphi(\alpha_0)\| + \|\varphi(\beta_0) - \varphi(\alpha_0)\|)dt$$
$$= M \cdot \epsilon_1,$$

where $M = 2(\|\varphi(\alpha_0)\| + \|\varphi(\beta_0) - \varphi(\alpha_0)\|)$. By the convergence of ℓ_α and r_β and by the uniform convexity of $(E, \|\cdot\|)$, we can take positive numbers ϵ_1, ϵ_2 such that there is a $\delta > 0$ and

(4.5.6)
$$\|f'(c) - \frac{z(c-u) + z(c+v)}{2}\| \leq \|f'(c) - z(c+v)\| - \delta$$

for all $0 < u \leq \epsilon_1$ and $0 < v \leq \epsilon_2$. For this ϵ_1, ϵ_2, we get

$$\int_c^{c+\epsilon_2} \|f'(t) - \varphi(t)\|dt \geq - \int_c^{c+\epsilon_2} \|f'(t) - f'(c)\| \, dt$$
$$+ \int_c^{c+\epsilon_2} \|f'(c) - \varphi(t)\| \, dt$$

71

$$\geq -\int_c^{c+\epsilon_2} \|f'(t) - f'(c)\| \, dt$$

$$+ \int_c^{c+\epsilon_2} \|f'(c) - z(c+\epsilon_2)\| \, dt,$$

and by (4.5.6),

$$\geq -\int_c^{c+\epsilon_2} \|f'(t) - f'(c)\| \, dt$$

$$+ \int_c^{c+\epsilon_2} \left(\|f'(c) - \frac{z(c-\epsilon_1) + z(c+\epsilon_2)}{2}\| + \delta \right) dt$$

$$(4.5.7) \quad \geq -2\int_c^{c+\epsilon_2} \|f'(t) - f'(c)\| \, dt$$

$$+ \int_c^{c+\epsilon_2} \|f'(t) - \frac{z(c-\epsilon_1) + z(c+\epsilon_2)}{2}\| \, dt + \epsilon_2 \delta.$$

Hence, if we consider an increasing function ψ such that

$$\psi(x) = \begin{cases} \varphi(x) & x \in (a, c - \epsilon_1) \cup (c + \epsilon_2, b) \\ \dfrac{z(c - \epsilon_1) + z(c + \epsilon_2)}{2} & x \in [c - \epsilon_1, c + \epsilon_2], \end{cases}$$

then, by (4.5.5) and (4.5.7),

$$\int_a^b \|f'(t) - \psi(t)\| dt = \int_a^{c-\epsilon_1} \|f'(t) - \varphi(t)\| \, dt$$

$$+ \int_{c-\epsilon_1}^{c+\epsilon_2} \|f'(t) - \psi(t)\| \, dt + \int_{c+\epsilon_2}^b \|f'(t) - \varphi(t)\| \, dt$$

$$(4.5.8) \qquad \leq \int_a^{c-\epsilon_1} \|f'(t) - \varphi(t)\| dt$$

$$+ \int_{c-\epsilon_1}^c \|f'(t) - \varphi(t)\| \, dt + M\epsilon_1 + \int_c^{c+\epsilon_2} \|f'(t) - \varphi(t)\| \, dt$$

$$+ 2\int_c^{c+\epsilon_2} \|f'(t) - f'(c)\| \, dt - \delta\epsilon_2 + \int_{c+\epsilon_2}^b \|f'(t) - \varphi(t)\| \, dt.$$

If we take positive numbers ϵ_1, ϵ_2 such that

$$2\|f'(x) - f'(c)\| < \delta \quad \text{for all } x \in [c, c + \epsilon_2],$$

and

$$\epsilon_1 M < \epsilon_2 \delta - 2\int_c^{c+\epsilon_2} \|f'(t) - f'(c)\| \, dt$$

by (4.5.8),

$$\int_a^b \|f'(t) - \psi(t)\| dt < \int_a^b \|f'(t) - \varphi(t)\| \, dt.$$

If we consider a convex function $h(x) = f(a) + \int_a^x \psi(t)dt$, then, by the above inequality, $\|f - h\|_v < \|f - \tilde{f}\|_v$. This contradicts the fact that $\tilde{f}$ is a best approximation to f. Thus φ is continuous on (a, b).

(uniqueness)   Suppose that f has distinct best approximations g_1, g_2 by K_2. If $\|f'(x) - g_1'(x)\| = \|f'(x) - g_2'(x)\|$ for some $x \in (a, b)$ and if $f'(x) - g_1'(x) \neq f'(x) - g_2'(x)$, then, from uniform convexity of $(E, \|\cdot\|)$, we have

$$(4.5.9) \quad \|f'(x) - \frac{g_1'(x) + g_2'(x)}{2}\| = \frac{1}{2}\|(f'(x) - g_1'(x)) + (f'(x) - g_2'(x))\|$$
$$< \|f'(x) - g_1'(x)\|.$$

Since g_1' and g_2' are continuous on (a, b) by the first half of this proof, for the other best approximation $(g_1 + g_2)/2$ to f by K_2, by (4.5.9), $v(f - (g_1 + g_2)/2) < v(f - g_1)$. This is a contradiction. Hence, whenever $\|f'(x) - g_1'(x)\| = \|f'(x) - g_2'(x)\|$ $x \in (a, b)$, $f'(x) - g_1'(x) = f'(x) - g_2'(x)$. Now we consider the following function h which is continuously differentiable on (a, b) and continuous on $[a, b]$:

$$h(a) = f(a)$$
$$h'(x) = g_i'(x) \quad x \in (a, b),$$

where $\|f'(x) - g_i'(x)\| = \min\{\|f'(x) - g_1'(x)\|, \|f'(x) - g_2'(x)\|\}$. Clearly $h'(x)$ is an increasing continuous function on (a, b) and integrable on $[a, b]$. By the distinction of g_1 and g_2, for the convex function $h(x) = f(a) + \int_a^x h'(t)dt$,

$$v(f - h) = \int_a^b \|f' - h'\|dt < \int_a^b \|f' - g_1'\|dt = v(f - g_1).$$

This contradicts the fact that g_1 is a best approximation to f. Consequently every $f \in BV_E - K_2$ has a unique best approximation by K_2, which is continuously diffrentiable on (a, b).

From an analogous method to the uniqueness part of the proof of Theorem 4.5.2, it follows that every sequence in bv_E has a unique best approximation by k_2.

Proposition 4.5.3. *Let $(E, \|\cdot\|)$ be a real reflexive strictly convex Banach space with a cone C. Suppose that $(E, \|\cdot\|)$ has $(C.1)$-property. Then any sequence $f \in bv_E - k_2$ has a unique best approximation $\tilde{f}$ by k_2.*

Let $(E, \|\cdot\|)$ be a real Hilbert space and let $<\cdot,\cdot>$ be the inner product on E. Let C be a cone satisfying

$$< u, v > \geq K\|u\|\|v\| \quad \text{for all } u, v \in C, \quad (***)$$

where K is a positive number independent on u and v. Then, for all $u, v \in C$, we have

$$\|u + v\|^2 = \|u\|^2 + 2 < u, v > + \|v\|^2$$
$$\geq \|u\|^2 + 2K\|u\|\|v\| + \|v\|^2$$
$$\geq (\|u\| + K\|v\|)^2,$$

because $0 < K \leq 1$. Hence,

$$(C.2) \qquad \|u + v\| \geq \|u\| + K\|v\| \quad \text{for all } u, v \in C.$$

Moreover, for all $u, v \in C$ with $\|u\| = 1$ and all $x \in E$ with $\|u + x\| \neq 0$, we see

$$< u + x + v, \frac{(u + x)}{\|u + x\|} > = \|u + x\| + \frac{< v, u + x >}{\|u + x\|}.$$

$$\geq \|u + x\| + \frac{\{K\|u\|\|v\| - \|x\|\|v\|\}}{\|u + x\|}.$$

If we put $\delta := \min\{K, 1/2\}$, then

$$(C.3) \qquad \|u + x + v\| \geq \|u + x\| \quad \text{for all } u, v \in C \text{ with } \|u\| = 1$$
$$\text{for all } x \in E \text{ with } \|x\| \leq \delta.$$

Using (C.2) and (C.3), we obtain

Theorem 4.5.4. *Let $(E, \|\cdot\|)$ be a real Hilbert space and let C be a cone with $(***)$-property. For a given f of $BV_E - K_i$, $i = 1, 2$,*

(1) if f is continuous, $P_i(f)$ consists of continuous functions

(2) if f is continuously differentiable, f has a unique best approximation by K_i, which is continuously differentiable on $[a, b]$.

Proof. By Theorem 4.4.4, Corollary 4.4.5 and Theorem 4.5.2, we only have to prove (2) when $i = 2$. Since, by Theorem 4.5.2, f has a unique best approximation $\tilde{f}$ by K_2 which is continuously differentiable on (a, b), we show that $\tilde{f}'$ is bounded on (a, b) and $\tilde{f}'(a + 0)$ and $\tilde{f}'(b - 0)$ exist.

Suppose to the contrary that $\tilde{f}'$ is unbounded on (a, b), say, $\lim_{x \to b-0} \|\tilde{f}'(x)\| = +\infty$. Since a subset $A = \tilde{f}'((a + b)/2) - \{f'(x) \mid x \in [a, b]\}$ is bounded, by (C.3), there exists an $\omega \in ((a + b)/2, b)$ such that

74

$$\|\tilde{f}'(x) - f'(x)\| = \|\tilde{f}'(\omega) - \tilde{f}'((a+b)/2) + \tilde{f}'((a+b)/2)$$

$$-f'(x) + \tilde{f}'(x) - \tilde{f}'(\omega)\|$$

(4.5.10) $$\geq \|\tilde{f}'(\omega) - \tilde{f}'((a+b)/2) + \tilde{f}'((a+b)/2) - f'(x)\|$$

$$= \|\tilde{f}'(\omega) - f'(x)\| \quad \text{for all } x \in (\omega, b).$$

If we consider an increasing function g on (a, b) such that

$$g(x) = \begin{cases} \tilde{f}'(x) & x \in (a, \omega) \\ \tilde{f}'(\omega) & x \in [\omega, b), \end{cases}$$

and set $k(x) = f(a) + \int_a^x g(t)dt$, then $\|f - k\|_v < \|f - \tilde{f}\|_v$. This is a contradiction. Hence $\tilde{f}'$ must be bounded on (a, b).

Next, suppose that, say, $\tilde{f}'(b - 0)$ does not exist. Since C has (C.1)-property, we can find a sequence $\{a_n\}$ which increasingly converges to b and

(4.5.11) $$\|\tilde{f}'(a_m) - \tilde{f}'(a_n)\| \geq \epsilon > 0 \quad \text{for } m, n \in N, m \neq n,$$

where ϵ is a positive number independent on m and n. For each $n \geq 3$, from (C.2) and (4.5.11), we get

(4.5.12) $$\|\tilde{f}'(a_n) - \tilde{f}'(a_1)\| = \|(\tilde{f}'(a_n) - \tilde{f}'(a_{n-1})) + (\tilde{f}'(a_{n-1}) -$$

$$\tilde{f}'(a_{n-2})) + \cdots + (\tilde{f}'(a_2) - \tilde{f}'(a_1))\|$$

$$\geq \|\tilde{f}'(a_n) - \tilde{f}'(a_{n-1})\| + K \sum_{i=1}^{n-2} \|\tilde{f}'(a_{i+1}) - \tilde{f}'(a_i)\|$$

$$\geq \|\tilde{f}'(a_n) - \tilde{f}'(a_{n-1})\| + \lambda(n-2)\epsilon.$$

But (4.5.12) contradicts the boundedness of $\tilde{f}'$ on (a, b). Hence, $\tilde{f}'(a + 0)$ and $\tilde{f}'(b - 0)$ exist. This completes the proof.

Remark 2. If a uniformly convex Banach space $(E, \|\cdot\|)$ with a cone C has (C.2) and (C.3)-property, then the same results as in Theorem 4.5.3 hold in $(E, \|\cdot\|)$.

Example. Let $(E, \|\cdot\|)$ be the real line $(R, |\cdot|)$ with $C = \{x \mid x \geq 0\}$. Let us consider the following continuously differentiable function $f(x)$ on $[-3, 3]$:

$$f(x) = \begin{cases} x^2 + 4x + 6 & x \in [-3, -2] \\ -x^2 - 4x - 2 & x \in [-2, -1] \\ x^2 & x \in [-1, 1] \\ -x^2 + 4x - 2 & x \in [1, 2] \\ x^2 - 4x + 6 & x \in [2, 3]. \end{cases}$$

By Theorem 4.5.4, f has a unique best approximation $\tilde{f} \in C^1[-3,3]$ by K_2. Since $\|f - g\|_v = \int_{-3}^{3} |f' - g'| dx$ for all $g \in C^1[-3,3]$, $\tilde{f}'$ is a unique best L^1-approximation to f' by $K_1 \cap C[-3,3]$. Using the method in Huotari, Legg, Meyerowitz and Townsend[10;p.135 and p.136], $\tilde{f}'$ is a step-like function $S(x)$ such that

$$S(x) = \begin{cases} 2x+4 & x \in [-3,-5/2] \\ -1 & x \in [-5/2,-1/2] \\ 2x & x \in [-1/2,1/2] \\ 1 & x \in [1/2,5/2] \\ 2x-4 & x \in [5/2,3]. \end{cases}$$

Hence, $\tilde{f}(x) = 3 + \int_{-3}^{x} S(t) dt$.

4.6 Problems

1. In 4.2, it is shown that $BV_E(\sigma(BV_E, F'))$ is a β-semi-Montel space if and only if $E(\tau)$ is semi-reflexive. In connection with this fact, we consider the following problem:

Let $E(\tau)$ be a semi-reflexive locally convex space . Find a class $\mathcal{C}$ of subspaces of F (F is the space appeared in 4.2.) such that $G(\sigma(G, F'))$ is β-semi-Montel for every $G \in \mathcal{C}$.

2. We restate problem 1 restricting it to real-valued sequence spaces on N.

Let ω be the space of all sequences and let φ be the subspace of ω which consists of finite sequences. Sequence spaces λ and η are said to form a *dual pair*, if (i) $< x,y >:= \sum_{i=1}^{\infty} x_i y_i$ exists for each $x = \{x_i\} \in \lambda$ and $y = \{y_i\} \in \eta$ and (ii) for each nonzero $x \in \lambda$ (resp. nonzero $y \in \eta$), there exists a $u \in \eta$ (resp. $v \in \lambda$) such that $< x,u > \neq 0$ (resp. $< v,y > \neq 0$).

Then we consider the following problem:

Characterize the class of sequence spaces λ such that λ and φ form a dual pair and $\lambda(\sigma(\lambda,\varphi))$ is β-Montel. (It is clear that $\lambda(\sigma(\lambda,\varphi))$ is infra-barrelled.)

For a sequence space λ, we define $\lambda^\beta = \{x \mid x \in \omega, < x,y > \text{ exists for each } y \in \lambda\}$. λ^β is called the β-*dual* of λ. If $\lambda = (\lambda^\beta)^\beta$, λ is said to be a β-*space* (see Kamthan and Gupta[14;p51 and p52]).

For example, ℓ^p, $1 \le p \le \infty$ and $bv = \{x \mid x = \{x_i\} \in \omega, \sum_{i=1}^{\infty} |x_{i+1} - x_i| \text{ exists }\}$ are

β-spaces. Hence, in connection with this problem, we state the following conjecture:

Let λ and φ be a dual pair. Then $\lambda(\sigma(\lambda, \varphi))$ is β-Montel if and only if λ is a β-space.

3. Let $(E, \|\cdot\|)$ be a uniformly convex Banach space with a cone C. By Corollary 4.4.5 and Remark 2 in 4.5, if $f \in BV_E - K_i$, $i = 1, 2$, is continuously differentiable, f has a unique best approximation by K_i. Similary, does the uniqueness of best approximations by K_i hold for any continuous function $f \in BV_E - K_i$?

4. Let $E(\tau)$ be a reflexive Banach space and let $K_{\mathcal{U}}$ be the set of all E-valued $\mathcal{U}$-convex functions (see p.7) in BV_E, where $\mathcal{U}$ denotes a Haar-like system in $C[a, b]$. Then, for a given Haar-like system $\mathcal{U}$, answer the corresponding problems of approximation by $K_{\mathcal{U}}$ to (P.1), (P.2) and (P.3) in 4.3. For example, in case $\mathcal{U} = \{1, x, \ldots, x^n\}$, what are the answers of these problems? This problem originates from the problem stated in Zwick[37].

Chapter 5 Approximation by Step Functions

5.1 Introduction

First we introduce an approximation problem of Jackson type which is treated in this chapter.

Let X be an infinite set and let $M(X)$ be the space of all real-valued bounded functions on X. $M(X)$ is endowed with the supremum norm $\| \cdot \|$, i.e., $\|f\| = \sup_{x \in X} |f(x)|$ for all $f \in M(X)$. For any subsets A, B of $M(X)$, $E_B(A) = \sup_{f \in A} \inf_{g \in B} \|f - g\|$. $\alpha(x, y)$ denotes a nonnegative bounded function on $X \times X$ and set $\mathcal{E}_\alpha = \{f \mid f \in M(X), |f(x) - f(y)| \leq \alpha(x, y) \text{ for all } x, y \in X\}$. For a given $\alpha(x, y)$ and a finite dimensional subspace G of $M(X)$, we consider the following problems:

(P.1) Estimate $E_G(\mathcal{E}_\alpha)$.

(P.2) Find approximating spaces G such that $E_G(\mathcal{E}_\alpha)$ is as small as possible.

Let Γ_α be the set of all pseudo-metrics $d(x, y)$ on X such that $d(x, y) \leq \alpha(x, y)$ for all $x, y \in X$. (A pseudo-metric $d(x, y)$ on X means a real-valued function on $X \times X$ satisfying, for all $x, y, z \in X$, (i) $d(x, y) \geq 0$, (ii) $d(x, x) = 0$, (iii) $d(x, y) = d(y, x)$, (iv) $d(x, y) + d(y, z) \geq d(x, z)$.)

Then we have

Proposition 5.1.1. *For a given* $\alpha(x, y)$*, there exists a pseudo-metric* $\rho_\alpha(x, y)$ *on* X *such that* $\mathcal{E}_\alpha = \mathcal{E}_{\rho_\alpha}$.

Proof. First we show that $\mathcal{E}_\alpha = \bigcup_{d \in \Gamma_\alpha} \mathcal{E}_d$. Since $d(x, y) \leq \alpha(x, y)$ for all $x, y \in X$, we easily see that $\mathcal{E}_\alpha \supset \bigcup_{d \in \Gamma_\alpha} \mathcal{E}_d$. For any function $g \in \mathcal{E}_\alpha$, when setting a pseudo-metric d on X such that $d(x, y) = |g(x) - g(y)|, x, y \in X$, we obtain $g \in \bigcup_{d \in \Gamma_\alpha} \mathcal{E}_d$.

If we put $\rho_\alpha(x, y) := \sup_{d \in \Gamma_\alpha} d(x, y)$, $x, y \in X$, then $\rho_\alpha(x, y)$ satisfies the conditions of pseudo-metrics on X and we observe $\mathcal{E}_\alpha = \mathcal{E}_{\rho_\alpha}$ without difficulty.

By Proposition 5.1.1, it does not lose its generality to consider the problems (P.1) and (P.2) under the condition that $\alpha(x, y)$ is a pseudo-metric on X.

When $\alpha(x, y)$ is a metric on X such that $(X, \alpha(x, y))$ is compact, Feinerman[1,2] and Feinerman and Newman[3] deeply studied these problems and showed that spaces of step functions can be considerably good approximating spaces.

In this chapter, based on their results, we study the generalized problems (5.1) and

78

(5.2). In 5.2, (P.1) and (P.2) are considered under the condition that G is a space of step functions. Specifically, we mention approximating spaces G such that $E_G(\mathcal{E}_\alpha)$ attains minimum among all 2-dimensional approximating spaces. In 5.3, we treat an approximation problem of set functions, which is reduced to the problems in 5.1. Some problems related to this topic are stated in 5.4.

5.2 Approximation by Step Functions

Before estimating $E_G(\mathcal{E}_\alpha)$, we give definitions and notations.

Definitions and Notations. Let $f(x,y)$ be a nonnegative bounded function on X satisfying $f(x,y) = f(y,x), x, y \in X$.

(1) The diameter of a subset A of X with f is denoted by $d_f(A)(= \sup_{x,y\in A} f(x,y))$.

(2) Let n be any positive integer and let A be any subset of X such that its cardinal number $|A| \geq n$. We write $\mathcal{D}_n(A)$ for the family of all n decompositions of A. For an n decomposition $\{B_i\}_{i=1}^n$ of A, we put $d_f(\{B_i\}_{i=1}^n) = \max\{d_f(B_1),\ldots,d_f(B_n)\}$. If $\{A_i\}_{i=1}^n$ is an n decompostion of X such that

$$d_f(\{A_i\}_{i=1}^n) = \inf_{\{B_1,\ldots,B_n\}\in\mathcal{D}_n(X)} d_f(\{B_i\}_{i=1}^n),$$

then we call $\{A_i\}_{i=1}^n$ an n decomposition with *MD-property* (minimum diameter).

(3) For any subset A of X, $\chi_A(x)$ denotes the characteristic function of A. In this chapter, step functions on X mean linear combinations of characteristic functions $\chi_{A_1},\ldots,\chi_{A_n}$, where $\{A_i\}_{i=1}^n \in \mathcal{D}_n(X)$. When G is a space spanned by characteristic functions $\chi_{A_1},\ldots,\chi_{A_n}$, $\{A_i\}_{i=1}^n \in \mathcal{D}_n(X)$, we use the notation $E_{(A_1,\ldots,A_n)}(\mathcal{E}_\alpha)$ instead of $E_G(\mathcal{E}_\alpha)$.

(4) For any positive integer n, the family of all sets of n distinct points in X is denoted by Δ_n, and we define

$$\epsilon_n(f) = \sup_{\{x_1,\ldots,x_{n+1}\}\in\Delta_{n+1}} \min_{i\neq j} f(x_i, x_j)$$

First we show lower bounds for $E_G(\mathcal{E}_\alpha)$. The following result follows from Theorem 5 in Chap. 8 in Feinerman and Newman[3].

79

Theorem 5.2.1. *Let G be an n-dimensional subspace of $M(X)$ and let $\alpha(x,y)$ be a nonnegative function in $M(X)$. Then*

$$E_G(\mathcal{E}_\alpha) \geq \frac{\epsilon_n(\rho_\alpha)}{2}.$$

Proof. For any $\epsilon > 0$, let $x_1, \ldots, x_{n+1}$ be $n+1$ points in X such that $\rho_\alpha(x_i, x_j) > \epsilon_n(\rho_\alpha) - \epsilon$ for $i \neq j$. Let $M = \{(u(x_1), \ldots, u(x_{n+1})) \mid u \in G\}$. Since M is a proper subspace of R^{n+1}, by Lemma 2.4.6, there exists a $(\sigma_1, \ldots, \sigma_{n+1})$ (Each $\sigma_i = 1$ or -1, $i = 1, \ldots, n+1$.) such that $(\sigma_1, \ldots, \sigma_{n+1}) \neq (s(u(x_1)), \ldots, s(u(x_{n+1})))$ for all $u \in G$. Let $T = \{x_i \mid \sigma_i = -1, 1 \leq i \leq n+1\}$ and we consider a function f on X such that

$$f(x) = \rho_\alpha(x, T) - \frac{\epsilon_n(\rho_\alpha)}{2} = \min_{x_i \in T} \rho_\alpha(x, x_i) - \frac{\epsilon_n(\rho_\alpha)}{2}.$$

For any $x, y \in X$, suppose that $\rho_\alpha(x, T) = \rho_\alpha(x, x_i)$ and $\rho_\alpha(y, T) = \rho_\alpha(y, x_j)$. Since $|f(x) - f(y)| \leq |\rho_\alpha(x, x_i) - \rho_\alpha(y, x_j)| \leq \rho_\alpha(x, y)$, f belongs to $\mathcal{E}_\alpha$. For any $u \in G$, there exists a point $x_i, 1 \leq i \leq n+1$ such that

$$u(x_i) \geq 0 \quad \text{and} \quad x_i \in T, \quad \text{or} \quad u(x_i) < 0 \quad \text{and} \quad x_i \notin T.$$

In any case, we have

$$\frac{\epsilon_n(\rho_\alpha)}{2} - \epsilon < \max_{1 \leq i \leq n+1} |f(x_i) - u(x_i)| \leq \|f - u\|.$$

Since ϵ is arbitrary, we obtain $E_G(\mathcal{E}_\alpha) \geq \frac{\epsilon_n(\rho_\alpha)}{2}$.

If approximating spaces are spaces of step functions, we state

Theorem 5.2.2. *Let $\alpha(x,y)$ be a nonnegative function in $M(X)$. For any $\{A_i\}_{i=1}^n \in \mathcal{D}_n(X)$, it holds that*

$$E_{(A_1, \ldots, A_n)}(\mathcal{E}_\alpha) = \frac{1}{2} d(\{A_i\}_{i=1}^n),$$

where $d(\{A_i\}_{i=1}^n)$ denotes the value with ρ_α.

Proof. By Proposition 5.1.1, it is sufficient to show that

$$E_{(A_1, \ldots, A_n)}(\mathcal{E}_{\rho_\alpha}) = \frac{1}{2} d(\{A_i\}_{i=1}^n).$$

For any $f \in \mathcal{E}_{\rho_\alpha}$, we approximate it using the step function $a_1^* \chi_{A_1}(x) + \cdots + a_n^* \chi_{A_n}(x)$, where $a_i^* = \frac{1}{2}(\sup_{x \in A_i} f(x) + \inf_{x \in A_i} f(x))$, $i = 1, \ldots, n$. Then we have, for any $x \in X$,

80

$$|f(x) - (a_1^* \chi_{A_1}(x) + \cdots + a_n^* \chi_{A_n}(x))|$$

$$\leq \frac{1}{2} \max_{1 \leq i \leq n} \{ \sup_{x \in A_i} f(x) - \inf_{x \in A_i} f(x) \}$$

$$\leq \frac{1}{2} d(\{A_i\}_{i=1}^n).$$

Hence, it holds that $E_{(A_1,\ldots,A_n)}(\mathcal{E}_{\rho_\alpha}) \leq \frac{1}{2} d(\{A_i\}_{i=1}^n)$.

On the other hand, suppose that $d(A_{i_0}) = d(\{A_i\}_{i=1}^n)$ for some i_0, $1 \leq i_0 \leq n$. For any $\epsilon > 0$, there are two points $s, t \in A_{i_0}$ such that $\rho_\alpha(s,t) > d(A_{i_0}) - \epsilon$. Then we consider a function $h(x) = \rho_\alpha(x,s)$, $x \in X$, which belongs to $\mathcal{E}_{\rho_\alpha}$. For any linear combination $a_1 \chi_{A_1}(x) + \cdots + a_n \chi_{A_n}(x)$, we have

$$\sup_{x \in A_{i_0}} |h(x) - (a_1 \chi_{A_1}(x) + \cdots + a_n \chi_{A_n}(x))|$$

$$= \sup_{x \in A_{i_0}} |h(x) - a_{i_0} \chi_{A_{i_0}}(x)|$$

$$\geq \frac{1}{2}(d(A_{i_0}) - \epsilon) > \frac{1}{2} d(\{A_i\}_{i=1}^n) - \epsilon.$$

Since ϵ is arbitrary, we get $E_{(A_1,\ldots,A_n)}(\mathcal{E}_\alpha) \geq \frac{1}{2} d(\{A_i\}_{i=1}^n)$. This completes the proof.

For an appropriate approximating space of step functions, we have the following estimation of $E_{(A_1,\ldots,A_n)}(\mathcal{E}_\alpha)$.

Theorem 5.2.3. *Let* $\alpha(x,y)$ *be a nonnegative function in* $M(X)$. *Then there exists an* n *decomposition* $\{A_i^*\}_{i=1}^n$ *of* X *such that*

$$(5.2.1) \qquad \frac{\epsilon_n(\rho_\alpha)}{2} \leq E_{(A_1^*,\ldots,A_n^*)}(\mathcal{E}_\alpha) \leq \frac{1}{2}(\epsilon_n(\rho_\alpha) + \epsilon_{n+1}(\rho_\alpha)).$$

Proof. By Theorem 5.2.1, we only have to show the latter half of (5.2.1). Since $\epsilon_n(\rho_\alpha) \geq \epsilon_{n+1}(\rho_\alpha)$, suppose that $\epsilon_n(\rho_\alpha) > \epsilon_{n+1}(\rho_\alpha)$. By this assumption, there exist $n + 1$ points $x_1, \ldots, x_{n+1}$ in X which satisfy

$$(5.2.2) \qquad \epsilon_n(\rho_\alpha) \geq \min_{i \neq j} \rho_\alpha(x_i, x_j) > \epsilon_{n+1}(\rho_\alpha)$$

$$(5.2.3) \qquad \rho_\alpha(x_n, x_{n+1}) = \min_{i \neq j} \rho_\alpha(x_i, x_j).$$

We examine the union $\cup_{i=1}^n V_{\epsilon_n(\rho_\alpha)}(x_i)$ of n closed balls, where $V_r(x_i) = \{x \mid \rho_\alpha(x_i, x) \leq r\}$. If $\cup_{i=1}^n V_{\epsilon_n(\rho_\alpha)}(x_i)$ is a proper subset of X, then we retake a point in $X - \cup_{i=1}^n V_{\epsilon_n(\rho_\alpha)}(x_i)$ as x_{n+1} and reorder $x_1, \ldots, x_{n+1}$ to satisfy the conditions (5.2.2), (5.2.3). By continuing this procedure at most n times, we have $n+1$ points $x_1^*, \ldots, x_{n+1}^*$ in X which satisfy the condition (5.2.2), (5.2.3) and

81

(5.2.4) $$\overset{n}{\underset{i=1}{\cup}} V_{\epsilon_n(\rho_\alpha)}(x_i^*) = X.$$

Putting the subset $A = \cup_{i=1}^n V_{\epsilon_{n+1}(\rho_\alpha)}(x_i^*)$ of X, we can see that, by (5.2.2) and (5.2.3), $X - A$ is not empty and that

(5.2.5) $$d(X - A) \leq \epsilon_{n+1}(\rho_\alpha),$$

where $d(X - A)$ denotes the diameter of $X - A$ with ρ_α. Provided that $d(X - A) > \epsilon_{n+1}(\rho_\alpha)$, there are $n + 2$ points $y_1, \ldots, y_{n+2}$ in X such that $\min_{i \neq j} \rho_\alpha(y_i, y_j) > \epsilon_{n+1}(\rho_\alpha)$. Now we set the n decomposition $\{A_i^*\}_{i=1}^n$ of X such that

$$A_1^* = V_{\epsilon_{n+1}(\rho_\alpha)}(x_1^*) \cup (V_{\epsilon_n(\rho_\alpha)}(x_1^*) \cap X - A),$$
$$\vdots$$
$$A_j^* = V_{\epsilon_{n+1}(\rho_\alpha)}(x_j^*) \cup (V_{\epsilon_n(\rho_\alpha)}(x_j^*) \cap X - A) - \cup_{k=1}^{j-1} A_k^*,$$
$$\vdots$$
$$A_n^* = X - \cup_{i=1}^{n-1} A_i^*.$$

Then, from (5.2.4) and (5.2.5), we have

$$d(A_i^*) \leq \epsilon_n(\rho_\alpha) + \epsilon_{n+1}(\rho_\alpha), \quad 1 \leq i \leq n.$$

Thus, by Theorem 5.2.2, $\{A_i^*\}_{i=1}^n$ is an n decomposition of X which we require. If $\epsilon_n(\rho_\alpha) = \epsilon_{n+1}(\rho_\alpha)$, from (5.2.4), we can find an n decomposition $\{A_i^*\}_{i=1}^n$ of X such that $d(A_i^*) \leq 2\epsilon_n(\rho_\alpha)$, $1 \leq i \leq n$. This completes the proof.

Suppose that $(E, \|\cdot\|)$ is a real normed space and B is an absolutely convex compact subset of E. If, for a point p in B, there exists an $\alpha > 0$ such that $p + \alpha B$ is contained in B, then we call this point p an *internal point* in B and denote the set of all internal points in B by B^i.(see p.176 in Köthe[6].)

When we consider the problem (P.1) on the compact metric space (B, ρ), where ρ is the metric such that $\rho(x, y) = \|x - y\|$, $x, y \in B$, a better estimation of $E_{(A_1, \ldots, A_n)}(\mathcal{E}_\alpha)$ than (5.2.1) is shown under the following condition.

Corollary 5.2.4. *Let B be an absolutely convex compact subset of real normed space and let (B, ρ) be the compact metric space stated above. For a given positive integer $n \geq 2$, assume that there exist n points $x_1, \ldots, x_n$ in B^i such that $\min_{i \neq j} \rho(x_i, x_j) \geq r_0 = \inf\{r \mid \cup_{i=1}^n V_r(x_i) := \cup_{i=1}^n \{x \mid x \in B, \rho(x, x_i) \leq r\} \supset B\}$. Then there is an n decomposition $\{B_i^*\}_{i=1}^n$ of B which satisfies*

(5.2.6) $$\frac{\epsilon_n(\rho)}{2} \leq E_{(B_1^*, \ldots, B_n^*)}(\mathcal{E}_\rho) \leq \frac{(r_0 + \epsilon_{n+1}(\rho))}{2} < \frac{(\epsilon_n(\rho) + \epsilon_{n+1}(\rho))}{2}.$$

Proof. We show the latter half of (5.2.6). From the compactness of (B, ρ), we can take a point y in B such that $\rho(y, x_i) \geq r_0$, $1 \leq i \leq n$. Since every x_i is contained in B^i, there exists an $\alpha > 0$ such that $x_i + \alpha B \subset B$ for $i = 1, \ldots, n$. Setting $u_i = x_i - y \in 2B$, $i = 1, \ldots, n$, we have $v_i = x_i + \frac{1}{2}\alpha u_i \in x_i + \alpha B \subset B$, $i = 1, \ldots, n$. Then, for $n + 1$ points $v_1, \ldots, v_n, y$ in B, we get

$$\rho(v_i, y) > r_0, \quad i = 1, \ldots, n$$

and for $i \neq j$,

$$\rho(v_i, v_j) > r_0.$$

Hence, we see $\epsilon_n(\rho) > r_0$.

If $r_0 \leq \epsilon_{n+1}(\rho)$, since $d(V_{r_0}(x_i)) \leq 2r_0$, $i = 1, \ldots, n$, the n decomposition $\{B_i^*\}_{i=1}^n$ of B such that $B_1^* = V_{r_0}(x_1)$, $\ldots$, $B_i^* = V_{r_0}(x_i) - \cup_{k=1}^{i-1} B_k^*$, $\ldots$, $B_n^* = V_{r_0}(x_n) - \cup_{k=1}^{n-1} B_k^*$ is suitable for our request.

If $r_0 > \epsilon_{n+1}(\rho)$ and if we put $C = \cup_{i=1}^n V_{\epsilon_{n+1}(\rho)}(x_i)$, then, by the definition of the values r_0 and $\epsilon_{n+1}(\rho)$, we can easily observe that $B - C$ is not empty and $d(B - C) \leq \epsilon_{n+1}(\rho)$. Thus, from an analogous method to the proof of Theorem 5.2.3, we have an n decomposition $\{B_i^*\}_{i=1}^n$ of B which satisfies (5.2.6). In any case, we obtain the conclusion.

Example. As an absolutely convex compact subset, let B be a closed cube of side 2 in R^3 endowed with the ℓ^2-norm $\|\cdot\|_2$. When $n = 8$, since B is divided into 8 cubes of side 1, we take the centers $x_1, \ldots, x_8$ of gravity of these small 8 cubes of side 1. For $x_1, \ldots, x_8$, we have

$$\min_{i \neq j} \|x_i - x_j\|_2 = 1 > r_0 = \frac{\sqrt{3}}{2}.$$

Hence, $x_1, \ldots, x_8$ satisfy the condition in Corollary 5.2.4.

Remark. (1) Theorem 5.2.3 and Corollary 5.2.4 give answers to the problems (P.1) and (P.2). Moreover these results show that spaces of step functions can be considerably good approximating spaces.

(2) Methods to obtain decompositions in Theorem 5.2.3 and Corollary 5.2.4 are closely related to those to obtain Dirichlet tilings with dot patterns. (See Appendix 1.)

(3) We can not show whether there exist internal points which satisfy the condition of Corollary 5.2.4 for every (B, ρ) and every n.

If there exists an n decomposition $\{A_i\}_{i=1}^n$ with MD-property, by Theorem 5.2.2,

$[\chi_{A_1}, \ldots, \chi_{A_n}]$ is a best approximating space among all n dimensional space of step functions. In fact, by Zorn's lemma, we can show the existence of an n decompostion with MD-property. We prove this in a more generalized form.

Theorem 5.2.5. *Let X be an infinite set and let $f(x, y)$ be a real-valued bounded function on $X \times X$ satisfying $f(x, y) = f(y, x)$, where $x, y \in X$. For a given positive integer n, there exists an n decomposition $\{C_i^*\}_{i=1}^n$ of X such that*

$$(5.2.7) \qquad d_f(\{C_i^*\}_{i=1}^n) = \inf_{\{A_i\}_{i=1}^n \in \mathcal{D}_n(X)} d_f(\{A_i\}_{i=1}^n).$$

Proof. For each ordinal number α, we inductively put the ordinal number ω_α such that $\omega_\alpha = \omega$ if $\alpha = 0$, and otherwise $\omega_\alpha :=$ the minimum ordinal number β satisfying $|\beta| > |\omega_{\alpha'}|$ for all $\alpha' < \alpha$. $|\beta|$ and $|\omega_{\alpha'}|$ denote their cardinal numbers. For each ordinal number α, let $P(\alpha)$ be the proposition which states that there exists an n decomposition with MD-property if $|X| = |\omega_\alpha|$.

We shall show that $P(0)$ is true. Let $\delta = \inf_{\{A_i\}_{i=1}^n \in \mathcal{D}_n(X)} d_f(\{A_i\}_{i=1}^n)$. Since $|X| = |\omega|$, we can assume that X is order isomorphic to ω. Let $X = \{x_i\}_{i<\omega}$ and set $X_k = \{x_0, \ldots, x_k\}$ for each $k (< \omega)$. Furthermore we set

$$\mathcal{D}_n := \{\{A_i\}_{i=1}^n \mid \{A_i\}_{i=1}^n \in \bigcup_{k \geq n} \mathcal{D}_n(X_k) \cup \mathcal{D}_n(X), d_f(\{A_i\}_{i=1}^n) \leq \delta\}$$

and induce the following order $\leq$ on $\mathcal{D}_n$:

$$\{A_i\}_{i=1}^n \leq \{B_i\}_{i=1}^n \quad \Longleftrightarrow \quad A_i \subset B_i, i = 1, \ldots, n.$$

Then $(\mathcal{D}_n, \leq)$ is an inductively ordered set. Because, for any linearly ordered subset $Y = \{\{A_i^\lambda\}_{i=1}^n \mid \lambda \in \Lambda\}$ of $(\mathcal{D}_n, \leq)$, when we set $A_i^* = \cup_{\lambda \in \Lambda} A_i^\lambda$, $i = 1, \ldots, n$, $\{A_i^*\}_{i=1}^n$ is an upper bound of Y. Hence, by Zorn's lemma, $(\mathcal{D}_n, \leq)$ has a maximal element $\{C_i^*\}_{i=1}^n$. Noting that, for each $k \geq n$, there exists $\{A_i\}_{i=1}^n \in \mathcal{D}_n(X_k)$ satisfying $\max\{d_f(A_1), \ldots, d_f(A_n)\} \leq \delta$, we must obtain $\cup_{i=1}^n C_i^* = X$. Thus $P(0)$ is true.

If $P(\beta)$ is true for all $\beta < \alpha$, by an analogous way to the above, we can prove the truth of $P(\alpha)$. Hence, by transfinite induction, the existence of n decomposition of X satisfying (5.2.7) is shown.

In particular, if $n = 2$, a constructive simple method to obtain a 2 decomposition with MD-property can be shown. We state this mehtod as a proposition.

Proposition 5.2.6. *Let X be an infinite set and let $f(x,y)$ be a real-valued bounded function on $X \times X$ which satisfies $f(x,y) = f(y,x)$, $x,y \in X$. Then, for f, there is a constructive method to obtain a 2 decomposition with MD-property.*

Proof. To make this proof simpler, we assume that X is a countable set $\{x_i\}_{i \in N}$. For the other cases, by the Well-Ordering Principle, X can be well ordered and the analogous method stated below can be applied. Let $\delta = \inf_{\{A,B\} \in \mathcal{D}_2(X)} d_f(\{A,B\})$ and, without loss of generality, we assume that $D(x) := \{y \mid y \in X, f(x,y) > \delta\} \neq \emptyset$ for all $x \in X$.

First, set $A_0^{(1)} := \emptyset$, $B_0^{(1)} := \emptyset$, $A_1^{(1)} := \{x_1\}$, $B_1^{(1)} := D(x_1)$ and for $k \geq 2$, set

$$A_k^{(1)} := \cup_{x \in B_{k-1}^{(1)}} D(x), \qquad B_k^{(1)} := \cup_{x \in A_k^{(1)}} D(x).$$

The sequences $\{A_k^{(1)}\}_{k=0}^{\infty}$ and $\{B_k^{(1)}\}_{k=0}^{\infty}$ of subsets of X are monotone increasing.

Now we shall show that $d_f(A_k^{(1)}) \leq \delta$ and $d_f(B_k^{(1)}) \leq \delta$ for all $k \in N$. It is easily seen that $d_f(A_1^{(1)}) \leq \delta$. Suppose to the contrary that $d_f(A_{k_0}^{(1)}) > \delta$ for some $k_0 \in N$, that is, there are two points, a and a', such that $f(a,a') > \delta$. Since $\{A_k^{(1)}\}$ is monotone increasing, let $A_p^{(1)}$ and $A_q^{(1)}$ be the subsets to which a and a' firstly belong, respectively. For convenience, we denote a and a' by a_p and a_q'. For the point a_p, if $a_p = x_1$, we put $S = \{x_1\}$. Otherwise, we take $2p - 1$ points $a_1 = x_1, a_2, \ldots, a_p$, $b_1, \ldots, b_{p-1}$ of X such that

(5.2.8) $\qquad a_i \in A_i^{(1)} - A_{i-1}^{(1)} \qquad\qquad i = 2, \ldots, p,$

(5.2.9) $\qquad b_j \in B_j^{(1)} - B_{j-1}^{(1)} \qquad\qquad j = 1, \ldots, p-1,$

(5.2.10) $\qquad f(a_i, b_i) > \delta, \quad f(a_{i+1}, b_i) > \delta \quad i = 1, \ldots, p-1,$

and we put $S = \{x_1, a_2, \ldots, a_p, b_1, \ldots, b_{p-1}\}$. By the same way, we can let $S' = \{x_1\}$ or $S' = \{x_1, a_2', \ldots, a_q', b_1', \ldots, b_{q-1}'\}$ of which points satisfy the conditions corresponding to (5.2.8) - (5.2.10) exist. If S and S' are ordered by

$$x_1 < b_1 < a_2 < \cdots < b_{p-1} < a_p$$

and

$$x_1 < b_1' < a_2' < \cdots < b_{q-1}' < a_q'$$

respectively, then, for $z = \max S \cap S'$, we set $C = \{x \mid x \geq z, x \in S\} \cup \{y \mid y \geq z, y \in S'\}$. Since C is a finite set of X which consists of odd elements, from (5.2.10) and $f(a_p, a_q') > \delta$, for any subset T of C, we have $\max\{d_f(T), d_f(C - T)\} > \delta$, which

85

contradicts the definition of δ. By an analogous reason, we also see that $d_f(B_k^{(1)}) \leq \delta$ for all $k \in N$. Hence, if we set $A_1^* = \cup_{k=1}^{\infty} A_k^{(1)}$ and $B_1^* = \cup_{k=1}^{\infty} B_k^{(1)}$, then $d_f(A_1^*) \leq \delta$ and $d_f(B_1^*) \leq \delta$ hold.

Next, let x_m be the point in $X-(A_1^* \cup B_1^*)$ with the smallest index. Then, by the same way as in $\{A_k^{(1)}\}$ and $\{B_k^{(1)}\}$, set $A_0^{(2)} := \emptyset$, $B_0^{(2)} := \emptyset$, $A_1^{(2)} := \{x_m\}$, $B_1^{(2)} := D(x_m)$, $\ldots, A_k^{(2)}, B_k^{(2)}, \ldots$ and set $A_2^* = \cup_{k=1}^{\infty} A_k^{(2)}$ and $B_2^* = \cup_{k=1}^{\infty} B_k^{(2)}$. By continuing these procedures, we obtain two sequences $\{A_k^*\}$ and $\{B_k^*\}$ of subsets of X such that

$$(5.2.11) \qquad A^* = \cup_{k=1}^{\infty} A_k^* \text{ and } B^* = \cup_{k=1}^{\infty} B_k^* \text{ is a two decompostion of } X,$$

$$(5.2.12) \qquad d_f(A_k^*) \leq \delta, \quad d_f(B_k^*) \leq \delta, \quad k \in N.$$

Since $d_f(x,y) \leq \delta$ for all $x \in A_k^* \cup B_k^*$ and all $y \in X - (A_k^* \cup B_k^*)$, by (5.2.11) and (5.2.12), $\{A^*, B^*\}$ is a 2 decompostion of X with MD-property. This completes the proof.

Here we mention very important results of 2-dimensional approximating spaces, which were obtained by Feinerman and Newman[3]. Let (X, ρ) be a compact metric space and let G be a 2-dimensional space of $M(X)$ spanned by $1, g$. Then Feinerman and Newman[3] showed the following:

(1) Suppose that g has only a finite number of values. Then there exists a 2 decomposition $\{A, B\}$ of X such that

$$E_{(A,B)}(\mathcal{E}_\rho) \leq E_G(\mathcal{E}_\rho).$$

(2) Let $\{g_n\}$ be a sequence of $M(X)$ such that g_n converges to g in the supremum norm, and let G_n be a space spanned by $1, g_n$. Then $E_{G_n}(\mathcal{E}_\rho)$ converges to $E_G(\mathcal{E}_\rho)$.

(3) Let $\mathcal{G}$ denote the set of all 2-dimensional spaces of $M(X)$. By (1) and (2),

$$\inf_{\{A,B\} \in \mathcal{D}_2(X)} E_{(A,B)}(\mathcal{E}_\rho) \leq \inf_{G \in \mathcal{G}} E_G(\mathcal{E}_\rho).$$

(4) For the metric ρ, there exists a 2 decomposition of X with MD-property.

When G is a 2-dimensional space of $M(X)$ which does not contain constants, and since $E_G(\mathcal{E}_\rho) = \infty$, G can not be a good approximating space. Hence, if $\{A^*, B^*\}$ is a 2 decomposition of X with MD-property, by (1) - (4), they had

$$E_{(A^*,B^*)}(\mathcal{E}_\rho) \leq E_G(\mathcal{E}_\rho) \quad \text{for all } G \in \mathcal{G}.$$

In fact, we see that (1) - (3) hold for an arbitrary pseudo-metric ρ on X by analogous proofs to those of (1) - (3). Thus, combining Proposition 5.1.1 with Proposition 5.2.6 or Theorem 5.2.5, we obtain

Theorem 5.2.7 *Let X be an infinite set and let $\alpha(x, y)$ be a real-valued nonnegative bounded function on $X \times X$. For the pseudo-metric ρ_α on X, let $\{A^*, B^*\}$ be a 2 decomposition of X with MD-property. Then*

$$E_{(A^*, B^*)}(\mathcal{E}_\alpha) \leq E_G(\mathcal{E}_\alpha) \quad \text{for all} \ \ G \in \mathcal{G}.$$

5.3 An Application to Approximation of Set Functions

In this section, we introduce an approximation problem which is reduced to the problems in 5.1.

Notations. The following notations are used:

(1) X is an infinite set.

(2) Σ is a σ-algebra of subsets of X.

(3) μ is a real-valued countably additive measure which is finite and nonatomic.

(4) $M(\Sigma)$ is the space of all real-valued functions on Σ. It is endowed with a norm $\|\cdot\|$ such that $\|f\| = \sup\limits_{A \in \Sigma} |f(A)|$ for all $f \in M(\Sigma)$.

(5) $\mathcal{E} = \{f \mid f \in M(\Sigma), |f(A) - f(B)| \leq \max\{\mu(A - B), \mu(B - A)\}$ for all $A, B \in \Sigma \}$.

Then we consider the following problem:

(P.3) Find n-dimensional approximating spaces G of $M(\Sigma)$ such that $E_G(\mathcal{E})$ is as small as possible.

Remark 1. (1) We easily see that (P.3) can be reduced to (P.2).

(2) Roughly speaking, $\mathcal{E}$ consists of set functions whose total variations are less than $\mu(X)$.

(3) Using a similar norm to the above, Shi[7] studies L-approximation problems in $C[a, b]$.

(4) Let $\rho(A, B) = \max\{\mu(A - B), \mu(B - A)\}$ for all $A, B \in \Sigma$. Since $A - C \subset (A - B) \cup (B - C)$ for all $A, B, C \in \Sigma$, it is observed that ρ is a pseudo-metric on Σ.

By the results of 5.2, step functions obtained by a 2 decomposition of (Σ, ρ) with

MD-property form a best approximating functions in (P.3). Now we mention 2 decompositions of (Σ, ρ) with MD-property.

Theorem 5.3.1. *Let (X, Σ, μ) and ρ be a measure space and a pseudo-metric on Σ in Notations, respectively. Set*

$\Sigma_1 = \{A \mid A \in \Sigma, 0 \le \mu(A) < \mu(X)/2\}$

$\Sigma_2 = \{A \mid A \in \Sigma, \mu(A) = \mu(X)/2\}$

$\Sigma_3 = \{A \mid A \in \Sigma, \mu(X)/2 < \mu(A) \le \mu(X)\}$.

Let $\{\mathcal{D}_1, \mathcal{D}_2\}$ be any 2 decomposition of (Σ, ρ) with MD-property such that $\emptyset \in \mathcal{D}_1$. Then the following statements hold:

(1) $\mathcal{D}_1 = \Sigma_1 \cup \Sigma_2'$ *and* $\mathcal{D}_2 = \Sigma_3 \cup \Sigma_2''$, *where Σ_2', Σ_2'' are disjoint subsets whose union is Σ_2.*

(2) $d_\rho(\{\mathcal{D}_1, \mathcal{D}_2\}) = \mu(X)/2 = \epsilon_2(\rho)$.

(3) $E_{(\mathcal{D}_1, \mathcal{D}_2)}(\mathcal{E}) = \dfrac{1}{4}\mu(X)$.

Proof. (1) and the first half of (2). Let $\{\mathcal{P}_1, \mathcal{P}_2\}$ $(\emptyset \in \mathcal{P}_1)$ be any 2 decomposition of (Σ, ρ) which is not represented as (1). Then, there exists an $A \in \mathcal{P}_1$ with $\mu(A) > \mu(X)/2$ or there exists a $B, X \in \mathcal{P}_2$ with $\mu(B) < \mu(X)/2$. In any case, we have $d_\rho(\{\mathcal{P}_1, \mathcal{P}_2\}) > \mu(X)/2$. Suppose that $\{\mathcal{Q}_1, \mathcal{Q}_2\}$ $(\emptyset \in \mathcal{Q}_1)$ is any 2 decomposition of (Σ, ρ) which is represented as (1). It is clear that $d_\rho(\mathcal{Q}_1) \le \mu(X)/2$. For any $A, B \in \mathcal{Q}_2$, if we set $A = A_0 + C$, $B = B_0 + C$, where $C = A \cap B$, then $\mu(A_0) + \mu(B_0) + \mu(C) \le \mu(X)$. From this, it immediately follows that $d_\rho(\mathcal{Q}_2) \le \mu(X)/2$. Hence, we obtain (1) and the first half of (2).

The latter half of (2). It is sufficient to show that $\mu(X)/2 = \epsilon_2(\rho)$. Let A, B, C be any three elements of Σ. Suppose that $\min\{\rho(A, B), \rho(B, C), \rho(C, A)\} > \mu(X)/2$. By this, there exist at least two subsets, for example A, B, of A, B, C such that $\mu(A) > \mu(X)/2$ and $\mu(B) > \mu(X)/2$. But, from the first half of this proof, we have $\rho(A, B) < \mu(X)/2$. This is a contradiction. Hence, $\min\{\rho(A, B), \rho(B, C), \rho(C, A)\} \le \mu(X)/2$ holds for any three elements of Σ. Since μ is nonatomic, we can take $\emptyset, A$ $(\mu(A) = \mu(X)/2)$ and X as three elements of Σ. Then, we get $\min\{\rho(\emptyset, A), \rho(\emptyset, X), \rho(A, X)\} = \mu(X)/2$.

(3). This follows from Theorem 5.2.2. This completes the proof.

Remark 2. (1) In (Σ, ρ), let A be an element in Σ with $\mu(A) = \mu(X)/2$. If we set $A_1 = \emptyset$, $A_2 = A$, $A_3 = X - A$ and $A_4 = X$, then $\epsilon_3(\rho) = \min_{i \ne j} \rho(A_i, A_j) = \mu(X)/2$. If $\{\mathcal{D}_1, \mathcal{D}_2\}$ is any 2 decomposition of (Σ, ρ) with MD-property such that $\emptyset \in \mathcal{D}_1$, then from Theorem 5.2.1 and 5.3.1, $[\chi_{\mathcal{D}_1}, \chi_{\mathcal{D}_2}]$ is a best approximating space of (P.3) among all at most 3-dimensional approximating spaces.

(2) When $n \geq 4$, we can not show $\epsilon_n(\rho)$ and concrete expressions of n decompositions of (Σ, ρ) with MD-property.

5.4 Problems

1. Let B be an absolutely convex compact subset of a real normed space $(E, \|\cdot\|)$ and let $\rho(x, y) = \|x - y\|$, $x, y \in B$. For a given positive integer $n(\geq 2)$, do there always exist n points $x_1, \ldots, x_n$ in B^i such that

$$\min_{i \neq j} \rho(x_i, x_j) \geq \inf\{r \mid \bigcup_{i=1}^n \{x \mid x \in B, \rho(x, x_i) \leq r\} \supset B\} \ ?$$

2. In 5. 2, a constructive method to obtain 2 decomposition with MD-property is shown. When $n \geq 3$, are there any constructive methods to decompose X into n subsets with MD-property ?

3. According to remarks in Feinerman[1], Theorem 5.2.7 does not always hold for $n(\geq 3)$ dimensional approximating spaces. Here we state a generalized version of Feinerman's conjecture(see [1]):

Let X be an infinite set and let $\alpha(x, y)$ be a real-valued nonnegative bounded function on $X \times X$. Then there exist n subsets $A_1^*, \ldots, A_n^*$ of X such that

$$E_{(A_1^*, \ldots, A_n^*)}(\mathcal{E}_\alpha) \leq E_G(\mathcal{E}_\alpha) \quad \text{for all } n\text{-dimensional subspaces } G \text{ of } M(X).$$

4. Let (Σ, ρ) be a pseudo-metric space in 5.3. When $n \geq 4$,

(1) estimate $\epsilon_n(\rho)$,

(2) show concrete expression n decomposition of (Σ, ρ) with MD-property,

(3) consider (P.3).

Appendix 1   A Visit to Dirichlet Tilings

How do you explain the meanings of the word "Tilings" and "Patterns" in regards to the plane? We can find the mathematical definitions of these words in the book by Grünbaum and Shephard (see p.16 and p.204 in [1]).

A *plane tiling* T is a countable family of closed subsets (= tiles) $\{T_1, T_2, \ldots\}$ which cover the plain without gaps or overlaps. To be more precise, the union of closed subsets is the whole plane and the interiors of closed subsets are pairwise disjoint. Any non-empty subset M is called a *motif*. A *pattern*(or *monomotif pattern*) P with motif M is a non-empty family $\{M_i\}_{i \in I}$ of subsets of the plane which satisfy the following conditions:

(1) The subsets M_i are pairwise disjoint.

(2) Each M_i is congruent to M.

(3) For each M_i, M_j, there is an isometry of the plane that maps P onto itself and M_i onto M_j.

Furthermore, in Grünbaum and Shephard [1], we can see profound studies of the classifications of tilings and patterns, and we see relations between tilings and patterns.

As an example of a natural way to associate a tiling with each pattern, the Dirichlet tiling is shown. Let $P = \{M_i\}_{i \in I}$ be a pattern. We associate a tile $T(M_i)$ with each M_i such that

$$T(M_i) = \{x \mid x \in R^2, d(x, M_i) \leq d(x, M_j) \text{ for all } j \in I\}.$$

$d(x, M_i)$ denotes the usual distance from M_i to x. In this case, the tiling $\{T(M_i)\}_{i \in I}$ is called the *Dirichlet tilings* associated with P.

In this appendix, we introduce a very simple method of designing tilings through a slight modification of Dirichlet tilings and we show some expressive tilings which we obtain by personal computers. Now we begin with our concrete trials.

Colored Dot Patterns (= C.D.P.) and Colored Dot Arrangements (= C.D.A.)

We use two simple patterns P_1, P_2 with a motif consisting of one point (dot pattern for short) and use two dot arrangements A_1, A_2. These dot patterns and dot arrangements consist of countable points in R^2. P_1 and P_2 consist of vertices of tiles of tilings $\{T_i\}$ such that T_i are congruent squares and congruent regular triangles, respectively. A_1 and A_2 consist of points which are radially arranged from a fixed point. Coordinate representations of points in P_1, P_2, A_1 and A_2 are as follows:

$P_1 = \{me_1 + ne_2 \mid n, m \in \mathbf{Z}\}$, where $e_1 = (1,0)$, $e_2 = (0,1)$.

$P_2 = \bigcup_{i=0}^{5} \{mx_i + nx_{i+1} \mid m, n \in \{0\} \cup \mathbf{N}\}$, where $x_i = (\cos i\pi/3, \sin i\pi/3)$, $0 \le i \le 6$.

$A_1 = \{c^k y_i \mid k \in \mathbf{Z}, i = 0, 1, 2, 3\}$, where $c > 1$, $y_i = (\cos i\pi/2, \sin i\pi/2)$, $0 \le i \le 3$.

$A_2 = \{c^k z_i \mid k \in \mathbf{Z}, i = 0, \ldots, 7\}$, where $c > 1$, $z_i = (\cos i\pi/4, \sin i\pi/4)$, $0 \le i \le 7$.

Fig.

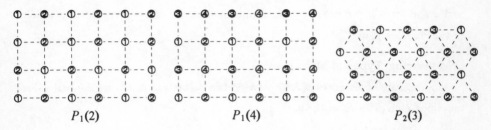

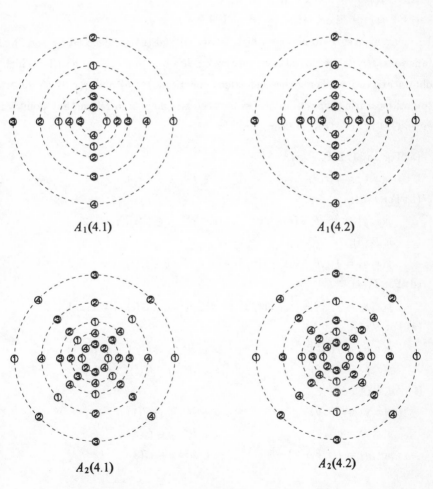

We regularly color points of these dot patterns and dot arrangements. Two or four colors are used for P_1, three colors for P_2 and four colors for A_1 and A_2. Each color is expressed as number(see Fig.). As for P_1 and P_2, the set of all points numbered i is denoted by $C(i)$. As for A_1 and A_2, we set

$D(i) = \{c^k \boldsymbol{y}_i \mid k \in \mathbf{Z}\}$, $0 \le i \le 3$, and $E(i) = \{c^k \boldsymbol{z}_i \mid k \in \mathbf{Z}\}$, $0 \le i \le 7$.

For any distinct points x and y in R^2, let r denote the usual distance between x and y and let θ, $0 \le \theta < 2\pi$, be the argument mod. 2π of $y - x$ regarded as a point in the complex plane.

Measuring Types of Distances

As a measure of distances $p(x, y)$ between two points x, y in R^2, we use real-valued functions f on $R^2 \times R^2$ such that

(i) $\quad f(x, x) = 0, \quad x \in R^2$,

(ii) $\quad f(x, y)$ is expressed by r and θ for $x \ne y$.

For convenience, the functions which satisfy (i) and (ii) are denoted by $f(r, \theta)$. The functions $f(r, \theta)$ are not always metrics on R^2, but this makes the tiles which we obtain expressive. For given real-valued functions $f(r, \theta)$ and $g(r, \theta)$ on $R^2 \times R^2$, several measuring types of distances between any point x in R^2 and any point y in a given C.D.P. or C.D.A. are considered.

(a) For any C.D.P. $P_i(j)$,

$\qquad p(x, y) = f(r, \theta) \qquad\qquad\qquad x \in R^2,\ y \in P_i(j).$

(b) For $P_1(2)$,

$\qquad p(x, y) = f(r, \theta + (i-1)\pi) \qquad x \in R^2,\ y \in C(i),\ i = 1, 2.$

(c) For $P_2(3)$,

$\qquad p(x, y) = f(r, \theta + 2(i-1)\pi/3) \quad x \in R^2,\ y \in C(i),\ i = 1, 2, 3.$

(d) For $P_1(4)$,

$\qquad p(x, y) = f(r, \theta + (i-1)\pi/2) \quad x \in R^2,\ y \in C(i),\ i = 1, 2, 3, 4.$

(e) For $P_1(4)$,

$\qquad p(x, y) = f(r, \theta) \qquad\qquad\qquad x \in R^2,\ y \in C(i),\ i = 1, 4,$

$\qquad p(x, y) = g(r, \theta) \qquad\qquad\qquad x \in R^2,\ y \in C(i),\ i = 2, 3.$

(f) For $P_1(4)$,

$\qquad p(x, y) = f(r, \theta) \qquad\qquad\qquad x \in R^2,\ y \in C(1)$

$\qquad p(x, y) = g(r, \theta) \qquad\qquad\qquad x \in R^2,\ y \in C(2)$

$\qquad p(x, y) = f(r, \theta + \pi) \qquad\qquad x \in R^2,\ y \in C(3)$

$$p(x,y) = g(r, \theta + \pi) \qquad\qquad x \in R^2, \ y \in C(4)$$

(g) For $A_1(4.1)$ or $A_1(4.2)$,

$$p(x,y) = f(r, \theta + i\pi/2) \qquad\qquad x \in R^2, \ y \in D(i), i = 0,1,2,3.$$

(h) For $A_2(4.1)$ or $A_2(4.2)$,

$$p(x,y) = f(r, \theta + i\pi/4) \qquad\qquad x \in R^2, \ y \in E(i), i = 0,\ldots,7$$

Designing Tilings

Let $P_i(j)$ or $A_i(j) = \{x_k\}_{k \in N}$ be a given C.D.P. or C.D.A. and let $p(x,y)$ be the function of a given measuring type which gives distances between points in R^2 and points in $P_i(j)$ or $A_i(j)$. Then, for each x_k, we design the tile $T(x_k) = \{x \mid x \in R^2, p(x,x_k) \le p(x,x_\ell) \text{ for all } \ell \in N\}$, $k \in N$, and the points of $T(x_k)$ are practically painted with the color of x_k. It is important to set functions $f(r,\theta)$ and $g(r,\theta)$ by which the family $\{T(x_k)\}_{k \in N}$ forms a tiling of R^2. Hence, we use continuous functions $f(r,\theta)$ and $g(r,\theta)$ such that, for each i and j, $i \ne j$, $\{x \mid x \in R^2, p(x,x_i) = p(x,x_j)\}$ has no interior points.

Following these procedures through the use of personal computers, we obtain various tilings. Some samples of them are as follows.

C.D.P.: $P_1(4)$ Measuring Type of Distance:(a)
$f(r,\theta) = r(0.5 \sin 2\pi\theta + 1)$

93

C.D.P.: $P_1(2)$ Measuring Type of Distance:(b)
$$f(r,\theta) = r(|\cos(\theta+1) + 1.5\cos 2\theta + 2\cos(3\theta-1)| + 1)$$

C.D.P.: $P_2(3)$ Measuring Type of Distance:(c)
$$f(r,\theta) = r(-0.01\theta(\theta-2)^2(\theta-4)^4(\theta-6)^2(\theta-2\pi) + 1)$$

C.D.P.: $P_1(4)$ Measuring Type of Distance:(d)
$$f(r,\theta) = r(|\cos(\theta+1) + \cos(2\theta+1) + \cos(3\theta+1)| + 1)$$

C.D.P.: $P_1(4)$ Measuring Type of Distance:(e)
$$f(r,\theta) = r(|\cos(\theta + 77) + \cos(2\theta + 77) + \cos(3\theta + 77)| + 1)$$
$$g(r,\theta) = r(0.01(\theta - 1.5)^2(\theta - 3.5)^2(\theta - 5)^2 + 1)$$

C.D.P.: $P_1(4)$ Measuring Type of Distance:(f)
$$f(r,\theta) = r(0.01(\theta - 1.5)^2(\theta - 3.5)^2(\theta - 5)^2 + 1)$$
$$g(r,\theta) = r(0.01\theta^2(\theta - 1)^2(\theta - 2)^2(\theta - 4.5)^2(\theta - 2\pi)^2 + 1)$$

C.D.A.: $A_1(4.1)$ Measuring Type of Distance:(g)
$$f(r,\theta) = r(-0.8\theta(\theta - 2)^2(\theta - 4)^4(\theta - 2\pi) + 1)$$

C.D.A.: $A_1(4.2)$ Measuring Type of Distance:(g)
$f(r,\theta) = r(0.5\sin 4\theta + 1)$

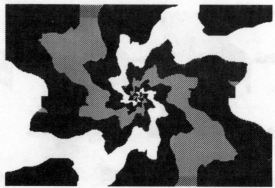

C.D.A.: $A_2(4.1)$ Measuring Type of Distance:(h)
$f(r,\theta) = r(0.5\sin \pi\theta + 1)$

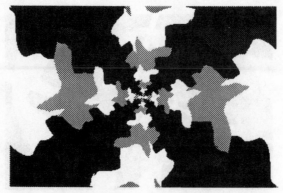

C.D.A.: $A_2(4.2)$ Measuring Type of Distance:(h)
$f(r,\theta) = r(|\cos(\theta+1) + \cos(2\theta+1) + \cos(3\theta+1)| + 1)$

If you adopt other types of C.D.P. or C.D.A., other types of functions $f(r, \theta)$ on $R^2 \times R^2$ and other types of measuring, more expressive tilings are easily obtained by the use of personal computers! I encourage you to try to design tilings. The program disk which we use here is sent to you if you want to try it and inform us to the following address:

Takakazu Yamamoto

Department of Business, Hamamatsu Junior College,
Hamamatu 430, Japan.

Appendix 2 On MD-Property

Let X be a set. Suppose that $f(x,y)$ is a not real-valued function but that it is a lattice-valued function on $X \times X$. In this appendix, we shall consider the existence of decompositions of X with MD-property. To facilitate our study, we prepare the following:

(1) Let L be a complete lattice with an order relation $\leq$, i.e., L is a partially ordered set such that every subset of L has a least upper bound and a greatest lower bound. For any subset A of L, we denote the least upper bound (resp. the greatest lower bound) of A by $\sup A$ (resp. $\inf A$).

(2) Let $f(x,y)$ be an L-valued bounded function on $X \times X$,i.e., there are two elements $u,v \in L$ such that $u \leq f(x,y) \leq v$ for all $x,y \in X$. Let A be a subset of X and let n be a positive integer. Then the L diameter of A with f is denoted by $d_f(A)$ $(= \sup_{x,y \in A} f(x,y))$. If $|A| \geq n$ and $\{A_i\}_{i=1}^n$ is an n decomposition of A, then we set $d_f(\{A_i\}_{i=1}^n) = \sup\{d_f(A_1), \ldots, d_f(A_n)\}$.

If $\{A_i\}_{i=1}^n$ is an n decompostion of X such that

$$d_f(\{A_i\}_{i=1}^n) = \inf_{\{B_i\}_{i=1}^n \in \mathcal{D}_n(X)} d_f(\{B_i\}_{i=1}^n),$$

then $\{A_i\}_{i=1}^n$ is called an n decomposition with MD-property.

Now we state

Theorem 1. *Let X be a set and let L be a complete lattice and let $f(x,y)$ be an L-valued bounded function on $X \times X$. For a given positive integer n, suppose that $\{\alpha_\lambda\}_{\lambda \in \Lambda}$ is a linearly ordered subset of L such that, for each α_λ, $\lambda \in \Lambda$, there is an n decomposition $\{A_i^\lambda\}_{i=1}^n$ of X with $d_f(\{A_i^\lambda\}_{i=1}^n) \leq \alpha_\lambda$. Then there exists an n decomposition $\{A_i^*\}_{i=1}^n$ of X such that $d_f(\{A_i^*\}_{i=1}^n) \leq \inf_{\lambda \in \Lambda} \alpha_\lambda$.*

Proof. For a given positive integer n, let $\{\alpha_\lambda\}_{\lambda \in \Lambda}$ be a linearly ordered subset of L satisfying the condition in Theorem 1.

Suppose that X is a finite set with $|X| = p(\geq n)$. Since $\{\alpha_\lambda\}_{\lambda \in \Lambda}$ is linearly ordered and $|\mathcal{D}_n(X)|$ is finite, there exist an n decomposition $\{A_i^*\}_{i=1}^n$ of X such that $d_f(\{A_i^*\}_{i=1}^n)$ is a lower bound of $\{\alpha_\lambda\}_{\lambda \in \Lambda}$.

Suppose that X is an infinite set. By a slight change of the proof of Theorem 5.2.5, we can prove this theorem. An outline of the proof follows.

For each ordinal number α, ω_α denotes the ordinal number defined in the proof of Theorem 5.2.5 and let $P(\alpha)$ be the proposition which states that Theorem 1 holds under $|X| = \omega_\alpha$. To show the truth of $P(0)$, let $X = \{x_i\}_{i<\omega}$ and set $X_k = \{x_0, \ldots, x_k\}$ for $(n \leq)k(< \omega)$ and

$$\mathcal{D}_n := \{\{A_i\}_{i=1}^n \mid \{A_i\}_{i=1}^n \in \bigcup_{k \geq n} \mathcal{D}_n(X_k) \cup \mathcal{D}_n(X), d_f(\{A_i\}_{i=1}^n) \leq \inf_{\lambda \in \Lambda} \alpha_\lambda\}$$

Using the order of L, we induce the following order $\leq$ on $\mathcal{D}_n$:

$$\{A_i\}_{i=1}^n \leq \{B_i\}_{i=1}^n \quad \Longleftrightarrow \quad A_i \subset B_i, i = 1, \ldots, n.$$

Since $(\mathcal{D}_n, \leq)$ is an inductively ordered set, by Zorn's lemma, $(\mathcal{D}_n, \leq)$ has a maximal element $\{A_i^*\}_{i=1}^n$. From the first half of this proof, we observe that, for each $k \geq n$, there exists $\{A_i\}_{i=1}^n \in \mathcal{D}_n(X_k)$ with $d_f(\{A_i\}_{i=1}^n) \leq \inf_{\lambda \in \Lambda} \alpha_\lambda$. Hence, we must obtain $\cup_{i=1}^n A_i^* = X$. Thus, $P(0)$ is true.

It is not difficult to see that Theorem 1 holds for any arbitrary set X through transfinite induction.

From Theorem 1, the following theorem immediately follows.

Theorem 2. *Let X be a set and let L be a complete lattice and let $f(x,y)$ be an L-valued bounded function on $X \times X$. For a given positive integer n with $|X| \geq n$, set $\alpha = \inf_{\{B_i\}_{i=1}^n \in \mathcal{D}_n(X)} d_f(\{B_i\}_{i=1}^n)$. Then there exists an n decomposition of X with MD-property if and only if there is a linear ordered set $\{\alpha_\lambda\}_{\lambda \in \Lambda}$ of L which satisfies the condition in Theorem 1 and whose $\inf_{\lambda \in \Lambda} \alpha_\lambda = \alpha$.*

Corollary 3. *Let X be a set and let L be a complete lattice and let n be a given positive integer with $|X| \geq n + 1$. For any L-valued bounded function $f(x,y)$ on $X \times X$, there always exists an n decompostion of X with MD-property if and only if L is linearly ordered.*

Proof. If L is linearly ordered, it follows from Theorem 2 that, for any L valued bounded function $f(x,y)$, there exists an n decompostion of X with MD-property.

Suppose that L is not linearly ordered. There are two elements $u, v \in L$ such that $u, v > w = \inf\{u, v\}$ and $u, v < z = \sup\{u, v\}$. Then we take $n+1$ points $x_1, \ldots, x_{n+1}$ in X and consider the L valued bounded function $f(x,y)$ on $X \times X$ such that

$$f(x,y) = f(y,x), \qquad x, y \in X$$
$$f(x_1, x_2) = u$$

99

$$f(x_i, x_j) = v, \qquad\qquad i \neq j, \{i,j\} \neq \{1,2\},$$
$$f(x, y) = w \qquad\qquad \text{otherwise.}$$

We observe that $d_f(\{A_i\}_{i=1}^n) = u, v$ or z for any $\{A_i\}_{i=1}^n \in \mathcal{D}_n(X)$. But $\inf_{\{A_i\}_{i=1}^n \in \mathcal{D}_n(X)}$ $d_f(\{A_i\}_{i=1}^n) = w$. This means that, for the L valued bounded function f, there does not exist an n decompostion of X with MD-property.

Remark. (1) Let X be a set and let L be a complete lattice. For a L-valued bounded function $f(x, y)$ on $X \times X$, suppose that there exists a 2 decomposition of X with MD-property. Then, from the same method in the proof of Proposition 5.2.5, we obtain a 2 decomposition of X with MD-property.

(2) When we consider countable decompositions instead of finite decompositions, the results stated above do not always hold. The existence of countable decompositions with MD-property depends on a set X and a lattice-valued function f on $X \times X$. Here we give two examples.

Let us consider $X = [0, 1]$(the real unit interval) and let $f(x, y) = |x - y|$, $x, y \in [0, 1]$(the usual metric on R). Since $[0, 1]$ is compact, we easily see that

$$\inf_{\{A_i\}_{i=1}^\infty \in \mathcal{D}_\omega([0,1])} d_f(\{A_i\}_{i=1}^\infty) = 0,$$

where $\mathcal{D}_\omega([0, 1])$ is the family of all countable decompositions of $[0, 1]$. Clearly, there does not exist a countable decompostion with MD-property.

Let us consider $X = [0, 1]$ and g such that $g(x, x) = 0$, $x \in [0, 1]$ and $g(x, y) = 1$ otherwise. Then, we observe that

$$\inf_{\{A_i\}_{i=1}^\infty \in \mathcal{D}_\omega([0,1])} d_f(\{A_i\}_{i=1}^\infty) = 1.$$

and every countable decomposition of $[0, 1]$ has MD-property.

(3) The readers can see detailed results about lattice in Birkhoff[1].

(4) The topic in this appendix is motivated by Horiuchi[2] and Kitahara and Tanaka[3].

References

Chapter1

1 D. A. Ault, F. R. Deutsch, P. D. Morris and J. E. Olson. Interpolating subspaces in approximation theory, *J. Approx. Theory* **3**(1970). 164 - 182.

2 E. W. Cheney, *Introduction to Approximation Theory*, McGraw Hill. New York. 1966, Reprint, Chelsea Publ. Co., New York, 1980.

3 P. J. Davis, *Interpolation and Approximation*, Dover Publ. Inc.. New York. 1975.

4 A. L. Garkavi, Haar's condition for systems of vector-valued functions. *Mathematical Notes* **45**(1989), 7 - 11.

5 S. Karlin and W. J. Studden, *Tchebycheff Systems:With Applications in Analysis and Statistics*, Wiley-Interscience Publ.. New York. 1966.

6 K. Kitahara, On Tchebysheff systems, *Proc. Amer. Math. Soc.* **105**(1989), 412 - 418.

7 G. G. Lorentz, *Approximation of Functions*, Chelsea Publ. Co.. New York. 1986.

8 G. Nürenberger, *Approximation by Spline Functions*. Springer-Verlag. Berlin Heidelberg, 1989.

9 A. W. Roberts and D. E. Varberg, *Convex Functions*, Academic Press. New York, 1973.

10 Y. G. Shi, The Chebyshev theory of a variation of L approximation, *J. Approx. Theory* **67**(1991), 239 - 251.

11 I. Singer, *Best Approximation in Normed Linear Spaces by Elements of Linear Subspaces*, Springer, New York, 1970.

12 S. Steckin, Approximation properties of sets in normed linear spaces, *Rev. Math. Pures Appl.* **8**(1963), 5 - 18.(Russian)

13 G. A. Watson, *Approximation Theory and Numerical Methods*. John Wilely and Sons Ltd, Chichester-New York-Brisbane-Tronto. 1980.

14 R. Zielke, *Discontinuous Čebyšev Systems*, Lecture Notes in Mathematics **707**, Springer-Verlag, Berlin-Heidelberg-New York. 1979.

15 D. Zwick, Existence of best n-convex approximations. *Proc. Amer. Math. Soc.* **97**(1986), 273 - 276.

Chapter 2

1 D. A. Ault, F. R. Deutsch, P. D. Morris and J. E. Olson. Interpolating subspaces in approximation theory, *J. Approx. Theory* **3**(1970), 164 - 182.

2 M. W. Bartelt, Weak Chebyshev sets and splines, *J. Approx. Theory* **14**(1975), 30 - 37.

3 J. Blatter, P. D. Morris and D. E. Wulbert, Continuity of the set-valued metric projection, *Math. Ann.* **178** (1968), 12 - 24.

4 E. W. Cheney, *Introduction to Approximation Theory*, McGraw Hill, New York, 1966, Reprint, Chelsea Publ. Co., New York, 1980.

5 F. Deutsch, G. Nürenberger and I. Singer, Weak Chebyshev subspaces and alternation, *Pacific J. Math.* **89** (1980), 9 - 31.

6 R. P. Feinerman and D. J. Newman, *Polynomial Approximation*, The Williams and Wilkins Company, Baltimore, 1974.

7 A. L. Garkavi, Almost Čebyšev systems of continuous functions, *Amer. Math. Soc. Transl.* **96**(1970), 177 - 187 (Translation of *Izv. Vysš. Učebn. Zaved. Matematika* **45**(1965), 36 - 44.).

8 A. Haar, Die Minkowskische Geometrie und die Annäherung an stetige Funktionen, *Math. Ann.* **78**(1918), 294 - 311.

9 D. C. Handscomb, D. F. Mayers and M. J. D. Powell, The general theory of linear approximation, in *Methods of Numerical Approximation*, Pergamon Press, New York, 1966.

10 E. Hewitt and K. Stromberg, *Real and Abstract Analysis*, Springer, New York, 1965.

11 R. C. Jones and L. A. Karlovitz, Equioscillation under nonuniqueness in the approximation of continuous functions, *J. Approx. Theory* **3**(1970), 138 - 145.

12 J. L. Kelly, *General Topology*, D. van Nostrand, New York, 1955.

13 K. Kitahara, A characterization of Chebyshev spaces, *Proc. Japan Acad.* **62** (1986), 375 - 378.

14 K. Kitahara, On Tchebysheff systems, *Proc. Amer. Math. Soc.* **105**(1989), 412 - 418.

15 K. Kitahara, A note on an infinite weak Tchebycheff system, to appear in J. Approx. Theory.

16 K. Kitahara, A characterization of almost and weak Chebyshev spaces, in preparation.

17 K. Kitahara and A. Ueda, A characterization of Tchebycheff systems, in preparation.

18 H. W. McLaughlin and K. B. Somers, Another characterization of Haar subspaces, *J. Approx. Theory* **14** (1975), 93 - 102.

19 G. Pólya and G. Szegö, *Problems and theorems in anlysis.* **I**, Springer- Verlag, New York, 1972.

20 L. Schumaker, *Spline functions: Basic theory*, Wilely-Interscience, New York, 1981.

21 M. Sommer, Weak Chebyshev spaces and best L^1-approximation, *J. Approx. Theory* **39**(1983), 54 - 71.

22 B. Stockenberg, On the number of zeros of functions in a weak Tchebyshev-space, *Math. Z.* **156**(1977), 49 - 57.

23 J. W. Young, General theory of approximation by functions involving a given number of arbitrary parameters, *Trans. Amer. Math. Soc.* **8**(1907), 331 - 344.

Chapter 3

1 E. W. Cheney and D. E. Wulbert, The existence and unicity of best approximations, *Math. Scand.* **24** (1969), 113 - 140.

2 G. Gierz and B. Shekhtman, On spaces with large Chebyshev subspaces, *J. Approx. Theory* **54** (1988), 155 - 161.

3 A. Haar, Die Minkowskische Geometrie und die Annäherung an stetige Funktionen, *Math. Ann.* **78**(1918), 294 - 311.

4 C. R. Hobby and C. R. Rice, A moment problem in L^1-approximation, *Proc. Amer. Math. Soc.* **65** (1965), 665 - 670.

5 D. Jackson, *The Theory of Approximation*, AMS Vol. XI, Colloq. Publ. Providence, Rhode Island, 1930.

6 S. Karlin and W. J. Studden, *Tchebycheff Systems: With Applications in Analysis and Statistics*, Wilely-Interscience Publ., New York, 1966.

7 K. Kitahara, On Tchebysheff systems, *Proc. Amer. Math. Soc.* **105**(1989), 412 - 418.

8 K. Kitahara, Best L^1-approximations, 505 - 512, *Progress in Approximation Theory* (P. Nevai and A. Pinkus Eds.), Academic Press, New York, 1991.

9 K. Kitahara, Representations of weak Tchebycheff systems, *Num. Func. Anal. Optim.* **14**(1993), 383 - 388.

10 M. G. Krein, The L-problem in abstract linear normed space, *Some Questions in the Theory of Moments* (N. Akiezer and M. Krein Eds.), Translations of Mathematical Monographs, Vol. 2, Amer. Math. Soc., Providence, Rhode Island, 1962.

11 B. R. Kripke and T. J. Rivlin, Approximation in the metric of $L^1(X, \mu)$, *Trans. Amer. Math. Soc.* **119**(1965), 101 - 122.

12 A. Kroó, On an L^1-approximation problem, *Proc. Amer. Math. Soc.* **94**(1985), 406 - 410.

13 A. Kroó, A general approach to the study of Chebyshev subspaces in L^1 approximation of continuous functions, *J. Approx. Theory* **51**(1987), 98 - 111.

14 A. Kroó, Best L^1-approximation with varying weights, *Proc. Amer. Math. Soc.* **99**(1987), 66 - 70.

15 A. Kroó, D. Schmidt and M. Sommer, On A-spaces and their relation to the Hobby-Rice theorem, *J. Approx. Theory* **68**(1992), 136 - 154.

16 W. Li, Weak Chebyshev subspaces and A-subspaces of $C[a, b]$, *Trans. Amer. Math. Soc.* **322**(1990), 583 - 591.

17 C. A. Micchelli, Best L^1-approximation by weak Chebyshev systems and the uniqueness of interpolating perfect splines, *J. Approx. Theory* **19**(1977), 1 - 14.

18 R. R. Phelps, Uniqueness of Hahn-Banach extensions and unique best approximation, *Trans. Amer. Math. Soc.* **95**(1960), 238 - 255.

19 R. R. Phelps, Cebysev subspaces of finite dimension in L_1, *Proc. Amer. Math. Soc.* **17**(1966), 646 - 652.

20 A. Pinkus, Unicity subspaces in L^1-approximation, *J. Approx. Theory* **48** (1986), 226 - 250.

21 A. Pinkus and B. Wajnryb, Necessity conditions for uniqueness in L^1-approximation, *J. Approx. Theory* **53**(1988), 54 - 66.

22 V. Ptak, On approximation of continuous functions in the metric $\int_a^b |x(t)|dt$, *Czechoslovak Math. J.* **8**(1958), 267 - 274.

23 M. A. Rutman, Integral representation of functions forming a Markov series, *Soviet Math. Dokl.* **164**(1965), 1340 - 1343.

24 D. Schmidt, A theorem on weighted L^1-approximation, *Proc. Amer. Math. Soc.* **101**(1987), 81 - 84.

25 F. Schwenker, Integral representation of normalized weak Markov systems, *J. Approx. Theory* **68**(1992), 1 - 24.

26 M. Sommer, *Approximation in Theorie und Praxis* (G. Meinardus, Ed.), Bibliographishes Institut, Mannheim, 1979.

27 M. Sommer, Weak Chebyshev spaces and best L^1-approximation, *J. Approx. Theory* **39**(1983), 54 - 71.

28 M. Sommer, Uniqueness of best L^1-approximation of continuous functions, *Delay Equations, Approximation and Applications* (G. Meinardus and G. Nürenberger, Eds.), Birkhäuser, Basel, 1985.

29 M. Sommer, Examples of unicity subspaces in L^1-approximation, *Num. Func. Anal. Optim.* **9**(1987), 131 - 146.

30 M. Sommer, Properties of unicity subspaces in L^1-approximation, *J. Approx. Theory* **52**(1988), 269 - 283.

31 H. Strauss, Best L^1-approximation, *J. Approx. Theory* **41**(1984), 297 - 308.

32 J. W. Young, General theory of approximation by functions involving a given number of arbitrary parameters, *Trans. Amer. Math. Soc.* **8**(1907), 331 - 344.

33 R. A. Zalik, Existence of Tchebycheff extensions, *J. Math. Anal. Appl.* **51** (1975), 68 - 75.

34 R. A. Zalik, On transforming a Tchebycheff system into a complete Tchebycheff system, *J. Approx. Theory* **20**(1977), 220 - 222.

35 R. A. Zalik, Integral representation of Tchebycheff systems, *Pacific. J. Math.* **68**(1977), 553 - 568.

36 R. A. Zalik, Integral representation and embedding of weak Markov systems, *J. Approx. Theory* **58**(1989), 1 - 11.

37 R. A. Zalik, Integral representation of Markov systems and the existence of adjoined functions for Haar spaces, *J. Approx. Theory* **65**(1991), 22 - 31.

38 R. A. Zalik, Nondegeneracy and integral representation of weak Markov systems, *J. Approx. Theory* **68**(1992), 30 - 42.

39 R. Zielke, On transforming a Tchebyshev-system into a Markov-system, *J. Approx. Theory* **9**(1973), 357 - 366.

40 R. Zielke, Alternation properties of Tchebyshev-systems and the existence of adjoined functions, *J. Approx. Theory* **10**(1974), 172 - 184.

41 R. Zielke, *Discontinuous Čebyšev Systems*, Lecture Notes in Mathematics **707**, Springer-Verlag, Berlin-Heidelberg-New York, 1979.

42 R. Zielke, Relative differentiability and integral representation of a class of weak Markov systems, *J. Approx. Theory* **44**(1985), 30 - 42.

43 D. Zwick, Characterization of WT-spaces whose derivatives form a WT-space, *J. Approx. Theory* **38**(1983), 188 - 191.

Chapter 4

1 V. Barbu and Th. Precupanu, *Convexity and Optimization in Banach Spaces* (revised and enlarged translation of convexitate si optimizare in spatii Banach), Sijthoff and noordhoff International Publishers, Netherlands, 1978.

2 R. B. Darst and S. Sahab, Approximation of continuous and quasi-continuous functions by monotone functions, *J. Approx. theory* **38**(1983), 9 - 27.

3 R. B. Darst and R. Huotari, Monotone L_1-approximation on the unit n-cube, *Proc. Amer. Math. Soc.* **95**(1985), 425 - 428.

4 R. B. Darst and R. Huotari, Best L_1-approximation of bounded approximately continuous functions on $[0, 1]$ by nondecreasing functions, *J. Approx. Theory* **43**(1985), 178 - 189.

5 E. Helly, Über lineare Funktionaloperationen, *Sber. Acad. Wiss. Wien.* **121** (1921), 265 - 297.

6 R. Huotari, Best L_1-approximation of quasi-continuous functions on $[0, 1]$ by nondecreasing functions, *J. Approx. Theory* **44**(1985), 221 - 229.

7 R. Huotari and D. Legg, Best monotone approximation in $L_1[0, 1]$, *Proc. Amer. Math. Soc.* **94**(1985), 279 - 282.

8 R. Huotari, A. D. Meyerowitz and M. Sheard, Best monotone approximations in $L_1[0, 1]$, *J. Approx. Theory* **47**(1986), 85 - 91.

9 R. Huotari and D. Legg, Monotone approximation in several variables, *J. Approx. Theory* **47**(1986), 219 - 227.

10 R. Huotari, D. Legg, A. D. Meyerowitz and D. Townsend, The natural best L_1-approximation by nondecreasing functions, *J. Approx. Theory* **52**(1988), 132 - 140.

11 R. Huotari and D. Zwick, Approximation in the mean by convex functions, *Numer. Funct. Anal. Optimiz.* **10**(1989), 489 - 498.

12 R. Huotari, D. Legg and D. Townsend, Existence of best n-convex approximants in L_1, *Approx. Theory Appl.* **5**(1989), 51 - 57.

13 R. Huotari, D. Legg and D. Townsend, Best L_1-approximation by convex functions of several variables, 475 - 481, *Progress in Approximation Theory* (P. Nevai and A. Pinkus, Eds.), Academic Press, New York, 1991.

14 P. K. Kamthan and M. Gupta, *Sequenece Spaces and Series*, Lecture Notes **65**, Marcel Dekker, New York, 1981.

15 S. Karlin and W. J. Studden, *Tchebycheff Systems: With Applications in Analysis and Statistics*, Wiley-Interscience, New York, 1966.

16 K. Kitahara, On two new classes of locally convex spaces, *Bull. Austral. Math. Soc.* **28**(1983), 383 - 392.

17 K. Kitahara, On a locally convex space admitting a fundamental sequence of strongly bounded subsets, *Publicacions Secció de Matemàtiques* **31**(1987), 85 - 109.

18 K. Kitahara, On the space of vector-valued functions of bounded variation, *Rocky Mountain J. Math.* **20**(1990), 165 - 171.

19 K. Kitahara and A. Nishi, Best approximations by vector-valued monotone increasing or convex functions, *J. Math. Anal. Appl.* **172**(1993), 166 - 178.

20 G. Köthe, *Topological Vector Spaces I* (translated by D. J. H. Garling, Die Grundlehren der mathematischen Wissenschaften **159**, Springer-Verlag, Berlin-Heidelberg-New York, 1969).

21 D. Legg and D. Townsend, Best monotone approximation in $L_\infty[0,1]$, *J. Approx. Theory* **42**(1984), 30 - 35.

22. A. Pinkus, Uniqueness in vector-valued approximation, *J. Approx. Theory* **73**(1993), 17 - 92.

23 A. W. Roberts and D. E. Varberg, *Convex Functions*, Academic Press, New York, 1973.

24 P. W. Smith and J. J. Swetits, Best approximation by monotone functions, *J. Approx. Theory* **49**(1987), 398 - 403.

25 J. J. Swetits and S. E. Weinstein, Construction of the best monotone approximation on $L_p[0,1]$, *J. Approx. Theory* **61**(1990), 118 - 130.

26 J. J. Swetits, S. E. Weinstein and Y. Xu, Approximation in $L_p[0,1]$ by n-convex functions, *Num. Func. Anal. Optim.* **11**(1990), 167 - 179.

27 V. A. Ubhaya, Isotone optimization. I, *J. Approx. Theory* **12**(1974), 146 - 159.

28 V. A. Ubhaya, Isotone optimization. II, *J. Approx. Theory* **12**(1974), 315 - 331.

29 V. A. Ubhaya, Uniform approximation by quasi-convex and convex functions, *J. Approx. Theory* **55**(1988), 326 - 336.

30 V. A. Ubhaya, L_p approximation by quasi-convex and convex functions, *J. Math. Anal. Appl.* **139**(1989), 574 - 585.

31 V. A. Ubhaya, L_p approximation by subsets of convex functions of sevaral variables, *J. Approx. Theory* **63**(1990), 144 - 155.

32 V. A. Ubhaya, S. E. Weinstein and Y. Xu, Best piecewise monotone unifrom approximation, *J. Approx. Theory* **63**(1990), 375 - 383.

33 D. Zwick, Existence of best n-convex approximations, *Proc. Amer. Math. Soc.* **97**(1986), 273 - 276.

34 D. Zwick, Best approximation by convex functions, *Amer. Math. Monthly* **94** (1987), 528 - 534.

35 D. Zwick, Characterizing shape preserving L_1-approximation, *Proc. Amer. Math. Soc.* **103**(1988), 1139 - 1146.

36 D. Zwick, Some aspects of best n-convex approximation, *J. Approx. Theory* **56**(1989), 30 - 40.

37 D. Zwick, Best L_1-approximation by generalized convex functions, *J. Approx. Theory* **59**(1989), 116 - 123.

Chapter5

1 R. P. Feinerman, A best two-dimensional space of approxmating functions, *J. Approx. Theory* **3**(1970), 50 - 58.

2 R. P. Feinerman, A best two-dimensional space of approxmating functions, II, *J. Approx. Theory* **4**(1971), 328 - 331.

3 R. P. Feinerman and D. J. Newman, *Polynomial Approximation*, The Williams and Wilkins Company, Baltimore, 1974.

4 K. Kitahara, A note on approximation by step functions, *Applied Math. Letters* **3**(1990), 23 - 25.

5 K. Kitahara and K. Tanaka, A two decomposition of a bounded metric space, *Applied Math. Letters* **4**(1991), 21 - 23.

6 G. Köthe, *Topological Vector Spaces I* (translated by D. J. H. Garling, Die Grundlehren der mathematischen Wissenschaften **159**, Springer-Verlag, Berlin-Heidelberg-New York, 1969).

7 Y. G. Shi, The Chebyshev theory of a variation of L approximation, *J. Approx. Theory* **67**(1991), 239 - 251.

Appendix 1

1 B. Grünbaum and G. C. Shephard, *Tilings and Patterns*, W. H. Freeman and Company, New York, 1989.

2 T. Yamamoto, T. Ohta and K. Kitahara, A visit to Dirichlet tilings, submitted.

Appendix 2

1 G. Birkhoff, *Lattice Theory*, A. M. S., Providence, 1948, Revised Edition, 1960.

2 K. Horiuchi, Private communication.

3 K. Kitahara and K. Tanaka, A two decomposition of bounded metric space, *Applied Math. Letters* **4**(1991), 21 - 23.

Index